Contents

Practical Guide to Inspection, Testing and Certification of
Electrical Installations

Practical Guide to Inspection, Testing and Certification of Electrical Installations

Conformity with the Regulations for Building

Christopher

ELSEVIER

PARIS • SAN DIEGO • SAN FRANCISCO • SINGAPORE • SYDNEY • TOKYO
Newnes is an imprint of Elsevier

Newnes

Newnes is an imprint of Elsevier
Linacre House, Jordan Hill, Oxford OX2 8DP, UK
30 Corporate Drive, Suite 400, Burlington, MA 01803, USA

First edition 2008

Notice
No responsibility is assumed by the publisher for any injury and/or damage to persons or
property as a matter of products liability, negligence or otherwise, or from any use or operation of
any methods, products, instructions or ideas contained in the material herein.

Disclaimer
This book draws on many sources. Some are facts, some hypotheses, some opinions. Most –
including many of my own statements – are mixtures. Even 'facts' are unavoidably selective
and can rarely be guaranteed. Despite careful checking, neither I nor my colleagues or publishers
can accept responsibility for any errors, misinformation or unsuitable advice. This also applies
to opinions – particularly on issues affecting health and safety. As any recommendations must
balance complex, often opposing, factors, not everyone will reach the same conclusions.
In this – as indeed in every issue this book touches on – every reader must make up her or his
mind, for which they alone must be responsible.

I offer the best advice I am capable of, but every circumstance is different. Anyone who acts on
this advice must make their own evaluation, and adapt it to their particular circumstances.

British Library Cataloguing in Publication Data
A catalogue record for this book is available from the British Library

Library of Congress Cataloging-in-Publication Data
A catalog record for this book is available from the Library of Congress

ISBN: 978-0-7506-8449-1

For information on all Newnes publications
visit our website at www.Elsevier.com

Printed and bound in Great Britain

08 09 10 11 12   10 9 8 7 6 5 4 3 2 1

Working together to grow
libraries in developing countries

www.elsevier.com | www.bookaid.org | www.sabre.org

ELSEVIER   BOOK AID
International   Sabre Foundation

Preface

Part P of the Building Regulations came into effect on the 1st of January 2005. This part of the Building Regulations requires that all electrical work carried out in domestic installations now has to be certificated. For work places, the Electricity at Work Regulations 1989 require that they provide a safe and well maintained electrical system. Certification and well kept records are a perfect way to confirm that every effort has been made to ensure that the system is, and remains safe.

I have written this book to assist electricians of all levels in carrying out the inspecting, testing and certification of all types of electrical installations. It will also be invaluable to **City and Guilds 2330** level 2 and 3 students, electricians studying for **City and Guilds 2391** parts 101 and 102 and all tradesmen who are required to comply with **Building Regulation part P**.

Step by step guidance and advice is given on how to carry out a detailed visual inspection during initial verifications, periodic inspections and certification.

Text and colour photographs of real, not simulated installations, are used to show the correct test instruments. Step by step instructions for how to carry out each test safely using different types of instruments are given along with an explanation as to why the tests are required.

In some photographs safety signs have been omitted for clarity. All test leads are to GS 38 where required. Some comments within this book are my own view and have been included to add a common sense approach to inspecting and testing.

Interpretation of test results is a vital part of the testing process. The correct selection and use of tables from BS 7671 and the On Site Guide are shown clearly. Any calculations required for correct interpretation or for the passing of exams are set out very simply. An in depth knowledge of maths is not required.

I have included questions and example scenarios, along with answers and completed documentation. These will assist electricians at all levels whether they need to pass an exam or complete the certification.

During my time lecturing I have been asked many questions by students who have become frustrated by being unable to reference definitive answers. Within this book I have tried to explain clearly and simply many of these difficult questions.

Chris Kitcher

Foreword

Christopher Kitcher is a very experienced electrician and teacher. I have worked with him for the last 11 years in both the college environment and '*on site*'. We are both examiners and also work for the City of Guilds on the 2391 qualification. Christopher now mainly works in education as well as writing books; this is his second book in the electrical installation sector, his first was the successful and very practical update to the WATKINS series.

Christopher's track-record speaks for it self; his work was instrumental in his college gaining the status of Centre of Vocational Excellence this is an accolade for training providers who gain a grade 1 or 2 during the joint OFSTED/ALI inspection.

Practical is again the keyword, as the book takes the reader from why to how in very clear steps. There are not only clear explanations, but photographs to guide the reader.

Whether an experienced electrician testing large industrial or commercial installations or a domestic installer altering or adding to installations, Chris has got it covered for you, generally using a language that installers use and not just technological terms that most installers find it difficult to understand. After all, we are electricians and not English language teachers, aren't we?

This new and exciting practical guide will support candidates who are looking to take exams such as the City and Guilds testing and inspection 2391 or 2, the EAL VRQ level 2 Domestic installer and the NIC Part P course.

> Gerry Papworth B.A. (Hons), MIET, Eng. Tech, LCGI
> Managing Director of Steve Willis Training (Portsmouth) Ltd
> Chief examiner for City & Guilds 2391

Acknowledgements

Writing this book has been a challenge which I have thoroughly enjoyed. It has been made a pleasurable experience because of the encouragement and generosity of many people and organisations. Particular thanks go to:

For the use of test equipment:
Phil Smith of Kewtech
Peter Halloway of Megger UK.

All at NIC certification for their help and assistance in allowing the reproduction of certificates and reports.

My granddaughter Heather Bates whose computer skills and patience saved the day on many occasions!

My colleagues at Central Sussex College for their valuable input, support and expertise:
Dave Chewter
Jason Hart
Lee Ashby
Andy Hay-Ellis
Jon Knight
Simon Nobbs

Brian Robinson for taking the photographs.

Finally, a special thanks is due to Central Sussex College for allowing me to use their facilities and equipment.

Chris Kitcher

Inspection and testing of electrical installations

Why inspect and test?

The Electricity at Work Regulations 1989 is a statutory document; it is a legal requirement that statutory regulations are complied with. Not to comply is a criminal offence and could result in a heavy fine and even imprisonment in extreme cases.

These regulations are required to ensure that places of work provide a safe, well-maintained electrical system. A simple way to provide this is to ensure that newly installed circuits and existing installations are tested on a regular basis. Electrical test certificates are used to record what has been done and confirm that the installation meets the required standard.

The British standard for electrical installations is BS 7671, the requirement for electrical installations. Within this standard, **Regulation 711-01-01** states that "every installation shall, during erection and on completion before being put into service be inspected and tested to verify, so far as reasonably practicable, that the requirements of the regulations have been met".

Regulation 731-01-02 states that "where required, periodic inspection and testing of every electrical installation shall be carried out in accordance with the requirements of Chapter 73".

Document P of the Building Regulations 2000 for Electrical Safety came into effect on 1 January 2005 and was amended in April 2006.

The purpose of this document is to ensure electrical safety in domestic electrical installations.

Section 1. Design, Installation, Inspection and Testing

This section of Part P is broken down into sub-sections.

General

This states that electrical work must comply with the Electricity at Work Regulations 1989 and that any installation or alteration to the main supply must be agreed with the electricity distributor.

Design and installation

This tells us that the work should comply with BS 7671 electrical wiring regulations.

Protection against flooding

The distributor must install the supply cut out in a safe place and take into account the risk of flooding. Compliance with The Electrical Safety, Quality and Continuity Regulations 2002 is required.

Accessibility

Part M of the building regulations must be complied with.

Inspection and testing before taking into service

This area is covered in detail throughout this book, it reminds us that the installation must be inspected and tested to verify that it is safe to put into service.

BS 7671 Installation certificates

This tells us that compliance with Part P can be demonstrated by the issue of a correct Electrical Installation Certificate and also what the certificate should cover. This is addressed later in this book.

Building regulation compliance certificates or notices for notifiable work

This tells us that the completion certificates issued by the local authorities, etc. are not the same as the certificates that comply with BS 7671. The completion certificates do not only cover Part P, but also shows compliance with all building regulations associated with the work which has been carried out.

Certification of notifiable work

This is covered in detail throughout this book.

Inspection and testing of non-notifiable work

This tells us that, even if the work is non-notifiable, it must be carried out to comply with BS 7671 and that certificates should be completed for the work.

Provision of information

Information should be provided for the installation to assist with the correct operation and maintenance. This information would comprise of certification, labels, instruction and plans.

Section 2. Extensions, Material alterations and material changes of use

This section is covered throughout this book, it basically tells us that certification is required, and that before any additions or alterations are made to an installation, an assessment of the existing installation should be made, to ensure that it is safe to add to.

Section 3. Information about other legislation

This covers the Electricity at Work Regulations 1989; Electrical Safety, Quality and Continuity Regulations 2002; functionality requirements.

The construction design and management regulations also state that adequate electrical inspection and tests are carried out on all new installations, those with electrical design information shall form a user's manual, which can be used to provide an up-to-date working record of the installation.

With the introduction of the 'Home Information Pack' (HIP) selling a property will eventually become very difficult if not impossible unless all of the relevant documentation is in place, this of course will include certification of electrical systems. Whilst, at the time of writing, this certification is not a requirement of the HIP, it is almost certain to become so in the future. Mortgage lenders and insurance companies are frequently asking for certification as part of the house buying/selling process. Owners of industrial and commercial properties could find that insurance is difficult to obtain, while most licensing bodies and local authorities are asking for electrical certification within their guidelines.

All of these regulations are under the umbrella of the Health and Safety at Work Act 1974. This clearly puts the legal responsibly of health and safety on all persons.

Compliance with Building Regulations Part P

Compliance with building regulations is a legal requirement and electrical work carried out in the domestic sector is now included in the building regulations; it is a criminal offence not to comply with the building regulations.

At the time of writing, there is no legal requirement to notify any work carried out in commercial or industrial buildings, although it should still be certificated for safety and record-keeping purposes.

Document P requires that most electrical work carried out in domestic premises is notified to the local authority building control. There are a few exceptions but the work must comply with BS 7671 Wiring Regulations. The exceptions are as follows:

Minor works carried out in areas that are not classed as special locations and therefore do not need notifying but would still need certifying
- Addition of socket outlets and fused spurs to an existing radial or ring circuit.
- Addition of a lighting point to an existing circuit.
- Installing or upgrading main or supplementary bonding.

Minor works carried out in the Special Locations as listed below – or in Kitchens (BS 7671 does not recognize a Kitchen as a special location. Document P does)
   Kitchens
   Locations containing bath tubs or shower basins
   Hot air saunas
   Electric floor or ceiling heating
   Garden lighting (*if fixed to a dwelling wall it is not deemed to come into the Special Location category*)
   Solar photovoltaic power supply systems

The work which could be carried out in these locations without notification but should still be certificated would be:

- Replacement of a single circuit which has been damaged

   Providing that the circuit follows the same route
   The cable used has the same current carrying capacity as the cable being replaced
   Circuit protective measures are not affected.

- Replacing accessories such as socket outlets, switches and ceiling roses.
- Re-fixing or replacing of enclosures and components.

All other work carried out in any areas of a domestic installation must be certificated and notified to the local authority building control, this can be carried out by various methods.

Earthing and bonding to comply with Part P

If a Minor Electrical Installation Works Certificate is necessary, there is no requirement to upgrade the existing earthing and bonding arrangements within an installation. Where the earthing and bonding do not comply with the latest edition of BS 7671, it should be recorded on the Minor Electrical Installation Works Certificate.

If an Electrical Installation Certificate is required, then the earthing arrangements must be upgraded to comply with the current edition of BS 7671.

Where the work is in the bathroom, or any areas that require supplementary bonding, then this must also be brought up to the current standard.

There is no requirement to upgrade supplementary bonding in an area where work is not to be carried out. There is also no requirement under Part P to certificate the upgrading of earthing and bonding to an installation.

Registered domestic installer

To become a registered domestic installer, it is necessary to become a member of one of the certification bodies which operate a domestic installer's scheme. This would require the person carrying out the work to prove competence in the type of work which is being carried out, and the ability to inspect, test and certificate the work which he/she has carried out. Competence is usually assessed by a site visit from an inspector employed by the chosen scheme provider.

There are two types of registration: (1) a person who needs to be able to carry out all types of electrical installation work in dwellings will need to register with an organization which runs a full scope scheme; (2) a person who needs to carry out electrical work associated with their main trade will need to register with an organization which runs a limited scope scheme. This scheme will enable a person to carry out electrical work which is related to the other work which is being carried out. An example of this would be where a person is a Kitchen fitter and needs to carry out electrical work which is required in the Kitchen. The installer would not be allowed to carry out electrical work in other parts of the dwelling unless that person was a member of a full scope scheme.

If the electrician is registered as a domestic installer, he or she must complete the correct certification and notify the scheme provider, who they are registered with, of the work which has been carried out. This must be done within 30 days. The scheme provider will both notify the local authority and the customer of the correct certification being given. An annual fee is usually required by the scheme provider, while a small fee is also payable for each job registered.

Unregistered competent person

If the work is carried out by a non-registered competent person who is capable of completing the correct certification, the local authority will need to be contacted before commencement of work, and the work will be carried out under a building notice. This will involve a fee being paid to the local authority and a visit or visits being made by a building inspector to inspect the work being carried out to ensure that it meets the required standard (*the cost of this will usually be far higher than that charged per notification by a scheme provider to a registered installer*). On satisfactory completion, and after the issue of the correct certification by the competent person, the building inspector will issue a completion certificate. The issue of a completion certificate by the local authority does not remove the responsibility for the work including guarantees from the non-registered competent person; the required certification must still be completed by the person who carried out or who is responsible for the work.

DIY installer

In cases where the work is carried out by a person who could not be deemed qualified (i.e. a DIY enthusiast), building control must be informed prior to work commencing, and on completion of the work to the building control officer's satisfaction, an inspection and test certificate must be issued. As a DIY installer would be unlikely to have the knowledge, experience or correct test equipment required to carry out the inspection, tests or completion of the certification, the services of a competent person would be required. The qualified person would in effect take responsibility for the new/altered work. For that reason, the qualified person would need to see the work at various stages of the installation to verify that the work and materials used comply with the required standards of the BS 7671 wiring regulations.

Summary

Currently, there is no requirement for any person carrying out electrical work in a domestic environment to be qualified in any way. The condition is that they must be competent; in other words, they must be in possession of the appropriate technical knowledge or experience to enable them to carry out the work safely.

There are Part P courses being provided by many training bodies, although it is not a requirement that you attend one of these courses or any other course which is being offered. However, it is impossible to become an electrician in 5 days.

The buildings control authorities must be informed of any electrical work that is to be carried on a domestic electrical installation other than very minor work, although even this work must be certificated.

Building control can be informed (*before commencing work*) by the use of a building notice, and this will involve a fee.

If your work involves a lot of domestic electrical work, then by far the best route would be to join one of the certification bodies. This would allow you to self-certificate your own work. When you join one of these organizations, you must be able to show that your work is up to a satisfactory standard and that you can complete the correct paperwork (*test certificates*). Whichever organization you choose to join, they will give you the correct advice on which training you require. A qualification is fine, but being able to carry out electrical work safely is far better.

2

Types of certification required for inspecting and testing of installations

Certification required for domestic installations (Part P)

The certification requirements for compliance with Part P are similar to the conditions for any other electrical installation.

It is a legal requirement to complete a Minor Electrical Installation Works Certificate (commonly called a 'Minor Works Certificate' or an 'Electrical Installation Certificate' for any electrical work being carried out on a domestic installation).

Minor Electrical Installation Works Certificate

This is a single document that must be issued when an alteration or addition is made to an existing circuit. A typical alteration that this certificate might be used for is the addition of a lighting point or socket outlet to an existing circuit. This certificate would be used for any installation regardless of whether it is domestic or not.

Part P, Domestic Electrical Installation Certificate

An Electrical Installation Certificate is required for:

- A new installation.
- When new circuits are installed.
- When a single circuit is installed.
- The changing of a consumer's unit.
- When a circuit is altered and the alteration requires the changing of the protective device.

This document is usually made up of three parts: (1) the Electrical Installation Certificate; (2) the Schedule of Inspections, and (3) the Schedule of Test Results (see Chapter 5). The format of these documents will differ slightly depending on who they are supplied by, but the content and legal requirement is the same.

Periodic inspection, testing and reporting

There is no requirement in Part P for periodic inspection, testing and reporting. However, if the replacement of a consumer's unit has been carried out, then the circuits which are reconnected should be inspected and tested to ensure that they are safe. This will, of course, require documentation: a Periodic Inspection Report, Schedule of Test Results and a Schedule of Inspection.

It is not a requirement of Part P that specific Part P certificates are used but you will find that many clients/customers prefer them.

The certificates produced by the IET (previously known as the IEE) are sufficient to comply with Part P and can be downloaded from www.theiet.org as described in the general certification section.

Some documents contain a **schedule of items tested**, which can also be found on the IET website. Although it is not a requirement that this document is completed, it is often useful as a checklist.

Certification required for the inspecting and testing of installations other than domestic

(Further explanation is provided for these documents in Chapter 5)

All of these certificates are readily available from many sources. The basic forms can be downloaded from the IET website which is www.theiet.org. Once on the site click on *publication*; next on *BS7671 Wiring Regulations*, then on *Forms for Electrical Contractor*s, and this will take you to all of the forms. If you scroll down the page a package is available that will allow you to fill in the forms before printing them.

The NICEIC have forms which can be purchased by non-members and most instrument manufacturers produce their own forms, which are also available from electrical wholesalers.

Minor Electrical Installation Works Certificate

This is a single document which should be issued if any alteration or addition is made to an existing circuit such as an additional lighting point or spurred socket outlet.

Electrical Installation Certificate

This certificate must be issued for a completely new installation or new circuit; this would include alterations to a circuit which would result in the changing of a protective device or the renewal of a distribution board.

The Electrical Installation Certificate must be accompanied by a Schedule of Test Results and a Schedule of Inspection. Without these two documents, the Electrical Installation Certificate is not valid. This certificate must not be issued until the installation complies with BS 7671.

An inspection and test which is carried out on a new installation to prove compliance is called an <u>initial verification</u>.

Initial verification inspection

The documentation which should be completed is the Electrical Installation Certificate; this must be accompanied by a Schedule of Test Results and a Schedule of Inspection.

The purpose of this inspection is to verify that the installed equipment complies with BS or BS EN standards; that it is correctly selected and erected to comply with BS 7671; and that it is not visibly damaged or defective so as to impair safety (*Regulation 712-01-01*).

When a new installation has been completed, it must be inspected and tested to ensure that it is safe to use. This process is known as the initial verification (Regulation 711-01-01). For safety reasons, the inspection process must precede testing.

Regulation 711-01-01 clearly tells us that the inspecting and testing process must be ongoing from the moment the electrical installation commences. In other words, if you are going to be responsible for completing the required

> You should never certificate any work which you have not seen during installation; once you sign the certificate you will be accepting a level of responsibility for it.

certification, you must visually inspect any parts of the installation which will eventually be covered up.

For this reason, by the time the installation is completed and ready for certification, a great deal of the installation will have already been visually inspected.

As an initial verification is ongoing from the commencement of the installation, much of the required inspecting and testing will be carried out during the installation, it is important that the whole range of inspection and tests are carried out on all circuits and outlets. Clearly it would not be sensible to complete the installation and then start dismantling it to check things like tight connections, fitting of earth sleeving and identification of conductors, etc.

There are many types of electrical installations and the requirements for them will vary from job to job. Where relevant, the following items should be inspected to ensure that they comply with BS 7671, during erection if possible:

* Have correct erection methods been used?
* Are diagrams and instructions available where required?
* Have warning and danger notices been fitted in the correct place?
* Is there suitable access to consumers' units and equipment?
* Is the equipment suitable for the environment in which it has been fixed?
* Have the correct type and size of protective devices been used?
* Have 30 mA residual current devices been fitted to circuits likely to supply portable equipment used outside? (*This could be socket outlets near windows*).
* Have 30 mA residual current devices been fitted where required to circuits supplying fixed current using equipment in zone 1 of the bathroom? (Regulation 601-09-02)
* Have 30 mA residual current devices been fitted where current using equipment other than fixed equipment has been installed in zone 3 of the bathroom? (Regulation 601-09-03)
* If the bedroom contains a shower have the socket outlets been protected by a 30mA residual current device? (Regulation 601-08-02)
* Are the isolators and switches fitted in the correct place?
* Could the installation be damaged by work being carried out on other services or by movement due to expansion of other services?
* Are bands 1 and band 2 circuits separated?
* Is there suitable protection against direct and indirect contact?
* Are fire barriers in place where required?
* Are the cables routed in safe zones? If not, are they protected against mechanical damage?

- Are the correct size cables being used, taking into account voltage drop and current carrying requirements?
- Are protective devices and single pole switches connected in the phase conductor?
- Are the circuits identified?
- Have the conductors been connected correctly?

This list is not exhaustive and, depending on the type of installation, other items may need to be inspected.

Initial verification testing

During the initial verification, each circuit must be tested. This will require the use of the correct type of testing equipment which is detailed later in this book.

For safety reasons, it is important that the testing procedure is carried out in the correct sequence, as stated in Guidance note 3 of BS 7671.

Sequence of tests

- Continuity of bonding conductors and circuit protective conductors.
- Continuity of ring final circuit conductors.
- Insulation resistance.
- Site applied insulation.
- Protection by separation of circuits.
- Protection by barriers and enclosures.
- Insulation of non-conducting floors.
- Dead polarity of each circuit.
- Live polarity of supply.
- Earth electrode resistance (Z_e).
- Earth fault loop impedance (Z_e) (Z_S).
- Prospective fault current (PFC).
- Functional testing.

Periodic inspection report

This document is for use in reporting and tracking the condition of an existing installation, and must be accompanied by a Schedule of Test Results and a Schedule of Inspection. Without these two documents the periodic test report is not valid.

A periodic inspection report would be carried out for many reasons. Examples are:

- The recommended due date
- Change of occupancy
- Change of use
- Change of ownership
- Insurance purposes
- Mortgage requirement
- Before additions or alterations
- After damage
- Client's request

Periodic inspection

A periodic inspection is carried out to ensure that the installation is safe and has not deteriorated.

The approach to this type of inspection is very different from that for an initial verification. It is vital that the original Electrical Installation Certificate, or past Periodic Inspection Reports, along with the Schedules of Test Results and the Schedules of Inspection, are available.

If this required documentation is not available, then the inspection and testing cannot proceed until a survey of the installation is carried out and fuse charts along with any other documentation that the inspector requires, is prepared.

The installation will have been used and the building is often occupied. It may possibly have had additions and alterations made to it. The type of use or even the environment could have changed from that which the installation was originally designed for.

Before commencing work the extent and limitation of the inspection must be agreed with the person ordering the work. A minimum of 10% of the installation should be inspected; this could increase, depending on any defects found.

Unlike an initial verification, the inspection should not be intrusive. Although covers will need to be removed in certain areas, it is not usually necessary to remove all accessories or carry out the full range of tests on every circuit. This will depend on what the inspector discovers as the inspection is carried out.

Visual inspection

What are we looking for during this inspection? In general terms we are inspecting the installation with regard to:

S afety
A ge
D eterioration
C orrosion
O verload
W ear and tear

An easy way to remember this is to use the acronym SADCOW.

Suitability and external influence should also be included. At this point it is a good idea to get from the client any documentation that is relevant to the installation. These documents could include:

- Plans
- Drawings
- Previous test results and certification
- Fuse charts

You should also make it clear that you will require access to all parts of the building and that the electricity supply will need to be turned off at some point. It is also a good idea to ask the client if they are aware of any alterations that have been carried out, as this information may be useful to you during inspection.

The visual inspection of any installation is as important as any testing that is carried out on an installation; if you are not familiar with the building it is also a good opportunity to find your way around first.

The first part of a visual inspection is to ensure that the system is safe to test and that you have enough information to be able to carry out the test safely. Generally, a good place to start would be the supply intake; this will give a reasonable indication of the age, type and size of the installation.

Things to look for at the supply intake <u>before</u> removal of any covers would be:

- The type of supply system – is it TT, TNS or TNCS?
- Is it old or modern?

- Are the conductors imperial or metric?
- What type of protection is there for the final circuits?
- Is documentation available for the original installation?
- Is the consumer's unit labelled correctly?
- Is the earthing conductor in place?
- What size is the earthing conductor?
- Is the earthing conductor green or green and yellow?
- Are all of the circuits in one consumer's unit or are there two or three units that need combining?
- Is there any evidence of equipotential bonding? Remember! It must start at the main earthing terminal.
- What size is the equipotential bonding? Is it large enough?
- Is there a residual current device (RCD)? If so has it a label attached? Is it a voltage or current operated type?
- Do the enclosures meet required IP codes? (Regulation 412-03-01)
- If alterations have been carried out is there documentation available for them, along with test results?
- Where alterations have been carried out since January 2005, has a warning notice been fitted on or near to the distribution board to indicate that new colours have been used? (Regulation 514-14-01)
- What size is the supply fuse? Is it large enough for the required load?
- Are the meter tails large enough?
- Are the seals broken on supply equipment? If they are it could indicate that the system has been tampered with since it was first installed and perhaps closer investigation is required.
- Have any alterations or additions been made?
- Would any alterations or additions affect the required disconnection time for the circuit concerned?

This list is not exhaustive and installation conditions may require more.

When the visual inspection of the supply intake area is complete, that is a good time to look around the building to make sure that there are no very obvious faults. All of this should be carried out before removal of any covers.

Things to look for:

- Are accessories fixed to the wall properly? Are they missing or damaged?
- Are the accessories old with wooden back plates?
- Are the socket outlets round pin or square? Is there a mixture of both?
- Have cables been installed in vulnerable situations?
- Have cables, enclosures and accessories been fixed securely?

- Have ceiling roses got perished flexes? (Particular attention should be given to the old braided and rubber type flexes.)
- Are any socket outlets likely to be used outside? If they are then they should be RCD protected. If they have been installed before the late 1990s, then it is not a requirement that they are, but an RCD should be listed as a recommendation.
- Are earthing clamps to BS 951 standards and correctly labelled?
- If gas, water is bonded using the same conductor, ensure that the conductor is continuous and not cut at the clamp.
- Is the supplementary bonding in place in bathroom? (See Figure 4 of the On-Site Guide.)
- Is the correct equipment for the correct zones in bath/shower room? (See 601 BS 7671)
- Has the bedroom had a shower installed? If so, are the socket outlets 3 metres from the shower and RCD protected?
- Is there any evidence of mutual detrimental influence; are there any cables fixed to water, gas or any other non-electrical services? (*The cables need to be far enough away to avoid damage if the non-electrical services are worked on.*)
- Are the cables of different voltage bands segregated? Low voltage, separated extra low voltage (SELV), telephone cables or television aerials should not be fixed together (*although they are permitted to cross*).

Whilst these items are being checked, look in any cupboards for sockets or lights. If your customer is uncomfortable with this it is vitally important that you document any areas that cannot be investigated in the extent and limitation section on the Periodic Inspection Report. During this purely visual part of the inspection you will gain some idea of the condition of the installation, and indeed any alterations which have been carried out by a qualified tradesman or by a cowboy/girl.

Clearly, if it is an old installation, an electrical installation certificate must be completed and some of the items listed above will apply. However, if it is a new installation, access to all areas must be secure; if this is not possible then the certificate should not be issued. Again, this list is not exhaustive but will not require removal of any fittings, etc.

Providing that you are happy that the installation is safe to tamper with, a more detailed visual inspection can be carried out and the dreaded but necessary form filling can be started.

Once again begin at the consumer unit. Before you start, this must be isolated. The Electricity at Work Regulations 1989 states that it is an offence to work live. Once you remove a cover you will be working live if you do not isolate it first. Having carried out the safe isolation procedure, remove the cover of the consumer unit.

- Your first impression will be important – has care been taken over the terminations of cables (*neat and not too much exposed conductor*)?
- Are all cables terminated and all connections tight (*no loose ends*)?
- Are there any signs of overheating?
- Is there a mixture of protective devices?
- Are there any rubber cables?
- Are there any damaged cables (*perished or cut*)?
- Have all circuits got Circuit Protective Conductors (CPCs)?
- Are all earthing conductors sleeved?
- On a photocopy of a Schedule of Test Results record circuits, protective devices and cable sizes.
- Look to see if the protective devices seem suitable for the size cables that they are protecting.
- Note any type D or 4 circuit breakers – these will require further investigation.
- Are all barriers in place?
- Have all of the circuit conductors been connected in sequence, with phase, neutral and CPC from circuit number 1 being in terminal number 1 – preferably the highest current nearest the main switch?
- Have any protective devices got multiple conductors in them, are they the correct size (*all the same*)?
- Is there only one set of tails or has another board been connected to the original board by joining at the terminals?

Having had a detailed look at the consumer unit, and with the installation still isolated, carry out a more detailed investigation of the rest of the installation.

It may be that you have agreed with your client that only 10% of the installation is to be inspected. This would mean 10% of each circuit. There would be little point in inspecting 10% of the circuits. If the period between inspections was 10 years it could be many years before a circuit was eventually inspected and the exercise would be pointless.

During your preliminary walk around, you will have identified any areas of immediate concern, and these must be addressed as your inspection progresses. There is no reason why you should not start your dead testing at this point, as you progress through your visual inspection.

On radial circuits this would be a good time to carry out CPC continuity, $R_1 + R_2$, insulation resistance and polarity tests as you work your way round. Start at circuit number 1 and work your way through the circuits one at a time.

But first what are you looking for? Let's look at a selection of circuits.

Shower circuit
- Is isolation provided, if so is it within prescribed zones?
- Has the correct size cable/protective device been selected?
- Is it bonded?
- Are connections tight?
- Has earth sleeving been fitted?
- Is the shower secure?
- Is there any evidence of water ingress?
- Is the shower in a bedroom?

Cooker circuit
- Is the switch within 2 metres of the cooker or hob?
- Has the cooker switch got a socket outlet? If so it requires a 0.4 second disconnection time.
- Green and yellow sleeving fitted.
- If it has a metal faceplate has it got an earth tail to the flush box?
- Is the cable the correct size for protective device?
- Are there any signs of overheating around the terminations?
- Is the cooker outlet too close to the sink? Building regulations require any outlets installed after January 2005 should be at least 300 mm from the sink.

Socket outlets
- Is there correct coordination between protective devices and conductors?
- Green and yellow sleeving fitted.
- Do any metal sockets have an earthing tail back to the socket box?
- Radial circuit not serving too large an area (*see Table 8A of the On-Site Guide*).
- Secure connections.
- Are cables throughout the circuit the same size?
- Are there any sockets outside? Are they waterproof? Are they 30 mA RCD protected?
- Are there any outlets in the bathroom? If there is, are they SELV?

- Are there socket outlets within 3 metres of a shower installed in a bedroom? If there is, are they 30 mA RCD protected?
- Will the protective device for the circuit provide 0.4 seconds disconnection time?

Fused connection units and other outlets
- As above but could be 5 second disconnection time.
- Does it supply fixed equipment in bathrooms? Are they in the correct zones?
- Do they require RCD protection? (*Regulation 601-09-02 and 601-09-03*)
- Permanently connected equipment must be protected locally by a plug or fused connection unit or comply with (*Regulation 476-03-04*).

Immersion heater circuits
- Is there correct coordination between the protective device and live conductors?
- Has the CPC been sleeved?
- Is the immersion the only equipment connected to this circuit? (*Any water heater with a capacity of 15 litres or more must have its own circuit.*) *On-Site Guide*, Appendix 8. Often you will find that the central heating controls are supplied through the immersion heater circuit.
- Is the immersion heater connected with heat resistant cord?
- The immersion heater switch should be a cord outlet type; not a socket outlet and plug.
- If the supplementary bonding for the bathroom is carried out in the cylinder cupboard, does the supplementary bonding include the immersion heater switch? (It should.)

Lighting circuits
- Is there correct coordination between the protective device and the live conductors?
- How many points are there on the circuit? A rating of 100 watts minimum must be allowed for each lighting outlet. Shaver points, clock points and bell transformers may be neglected for the purpose of load calculation. As a general rule, ten outlets per circuit is about right. Also remember that fluorescent fittings and discharge lamps are rated by their output, and the output must be multiplied by a factor of 1.8 if exact information is not available (*Table 1A of the On-Site Guide*).
- Are the switch returns colour identified at both ends?
- Have the switch drops got CPCs? If they have, are they sleeved with green and yellow?
- Are the CPCs correctly terminated?

- Are the switch boxes made of box wood or metal?
- Are ceiling roses suitable for the mass hanging from them?
- Only one flexible cord should come out of each ceiling rose unless they are designed for multiple cords.
- Light fittings in bathrooms must be suitable for the zones in which they are fitted.
- Circuits supplying luminaries fitted outside must have a 0.4 second disconnection time (*Regulation 471-08-03*).
- Is the phase conductor to ES lampholders connected to the centre pin? This does not apply to E14 and E27 lampholders (*Regulation 553-03-04*).

Three phase circuit/systems

These circuits should be inspected for the same defects that you could find in other circuits. In addition to this:

- Are warning labels fitted where the voltage will be higher than expected? For example, a lighting switch with two phases in it, or perhaps where sockets close to each other are on different phases.
- Are conductors in the correct sequence?
- Remember PFC should be double the phase to neutral fault current.

Occasionally other types of circuit will be found, but the same type of inspection should be carried out using common sense.

> Always remember that the reason for this inspection is to ensure safety

Periodic testing

The level of testing will usually be far less for periodic testing than it is for initial verification; this is providing that previous test results are available. If they are not, then it will be necessary for the full survey and the complete range of tests to be carried out on the installation, to provide a comprehensive set of results.

The level of testing will depend largely on what the inspector discovers during the visual inspection, and the value of results obtained while carrying out sample testing. If any tests show significantly different results, then further testing may be required.

In some cases, up to 100% of the installation will need to be tested. Periodic testing can be dangerous, and due consideration should be given to safety.

Persons carrying out the testing must be competent and experienced in the type of installation being tested and the test instruments being used.

There is no set sequence for the testing which may be required for the completion of the periodic inspection report. The sequence and type of tests which are to be carried out are left to the person carrying out the test to decide upon. Where tests are required, the recommendations for these tests would be:

	Recommended tests
Continuity of protective conductors	Between the distribution board earth terminal and exposed conductive parts of current using equipment. Earth terminals of socket outlets (test to the fixing screw of outlet for convenience).
Continuity of bonding conductors	All main bonding and supplementary bonding conductors.
Ring circuit continuity	Only required where alterations or additions have been made to the ring circuit.
Insulation resistance	Only between live conductors joined and earthed. Or between live conductors with the functional switch open if testing lighting circuit.
Polarity	Live polarity tested at the origin of the installation. Socket outlets. At the end of radial circuits. Distribution boards.
Earth electrode resistance	Isolate installation and remove earthing conductor to avoid parallel paths.
Earth fault loop impedance	At the origin of the installation for Z_e. Distribution boards for the Z_e of that board. Socket outlets and at the end of radial circuits for Z_s.
Functional tests	RCD tests and manual operation of isolators, protective devices and switches.

Voltage drop in conductors

It is part of the inspection process to ensure that installed conductors have been correctly selected for current carrying capacity and voltage drop. To check the suitability of the current carrying capacity it is simply a matter of looking at the installation method, and then checking on the current carrying capacity tables for the cable in Appendix 4 of BS 7671.

To ensure that the cable meets the voltage drop requirements is slightly more complex. A simple method is to measure the voltage at the origin of the circuit, and then measure the voltage at the end of the circuit with the load connected and switched on. The difference between the two measurements will be the volt drop.

If the first method is impractical, then a resistance test should be carried out between the phase and neutral of the circuit. This test is carried out using the same method as the $R_1 + R_2$ test although, instead of the test being between phase and CPC, it is between the phase and neutral for the circuit. Once the resistance $R_1 + R_n$ of the circuit has been measured it should be multiplied by the current that will flow in the circuit. This will give you the volt drop for the circuit.

Example

A circuit is wired in $2.5 \, \text{mm}^2$ and is 25 metres in length. The current in the circuit is 18 amps.

The measured value of resistance is 0.37 Ω

$$Voltage \; drop = I \times R = V$$
$$18 \times 0.37 = 6.66 \; volts.$$

This is the voltage drop for the circuit.

3

Testing of electrical installations

Safe isolation

It cannot be over-emphasized how important it is that isolation of electrical circuits is carried out in a set sequence, and that this sequence is repeated each time a circuit or complete installation is to be isolated.

If the same procedure is followed each time isolation is carried out, it will soon become a habit, which can only be a good thing as it may save your life.

It is vital that the correct test equipment is used for isolation and that it complies with the Health and Safety Executive document GS 38. This document gives guidance on the use of test equipment, particularly leads and probes.

The GS 38 document is not a statutory document but if the guidance given in the document is followed, it will normally be enough to comply with the Health and Safety at Work Act 1974, the Electricity at Work Regulations 1989 and any other statutory requirements that may apply. The items of equipment that should be available to persons carrying out the safe isolation procedure are:

• A proving unit

- An approval voltage
  indicator (left) and
  test Lamp (right)

- Warning notices

- Locking devices

Another useful piece of equipment is:

- An R_1 and R_2 box. This will not only be useful for the safe isolation of socket outlets, it can also be used for ring circuit testing and the $R_1 + R_2$ testing of radial circuits incorporating a socket or socket outlets without having to remove them from the wall.

R$_1$ and R$_2$ box
Neutral   Earth   (Live)
phase

The leads should be:

- Flexible and long enough, but not too long.
- Insulated to suit the voltage at which they are to be used.
- Coloured where it is necessary to identify one lead from the other.
- Undamaged and sheathed to protect them against mechanical damage.

The probes should:

- Have a maximum of 4 mm exposed tip (preferably 2 mm).
- Be fused at 500 mA or have current limiting resistors.
- Have finger guards (*to stop fingers slipping on to live terminals*).
- Be colour identified.

Isolation procedure

It is very important to ensure that the circuit that you want to isolate is live before you start. To check this, a voltage indicator/test lamp or a piece of equipment that is already connected to the circuit should be used. If it appears that the circuit is already dead, you need to know why.

- Is somebody else working on it?
- Is the circuit faulty?
- Is it connected?
- Has there been a power cut?

You must make absolutely certain that you and you alone are in control of the circuit to be worked on. Providing the circuit is live you can proceed as follows:

STEP 1

Ensure voltage indicator/test lamp is working correctly.

Voltage lights lit

STEP 2

Test between all live conductors and live conductors and earth.

Voltage lights lit

STEP 3

Locate the point of isolation. Isolate and lock off.

Place warning notice (**DANGER ELECTRICIAN AT WORK**) at the point of isolation.

Isolate and lock off   Place warning notice

STEP 4

Test circuit to prove that it is the correct circuit that you have isolated.

No voltage lights lit

Be careful! Most test lamps will trip an RCD when testing between live and earth, it is better to use an approved voltage indicator to GS 38 as most of these do not trip RCDs

STEP 5

Check that the voltage indicator is working by testing it on a proving unit or a known live supply.

Voltage lights lit

When carrying out the safe isolation procedure never assume anything, always follow the same procedure

It is now safe to begin work.

If the circuit which has been isolated is going to be disconnected at the consumer's unit or distribution board, **REMEMBER** the distribution board should also be isolated. The Electricity at Work Regulations 1989 do not permit live working.

Testing for continuity of protective conductors

Main equipotential bonding

This test is carried out to ensure that the equipotential bonding conductors are unbroken, and have a resistance low enough to ensure that, under fault conditions, a dangerous potential will not occur between earthed metalwork (*exposed conductive parts*) and other metalwork (*extraneous conductive parts*) in a building.

It is not the purpose of this test to ensure a good earth fault path but to ensure that, in the event of a fault, all exposed and extraneous conductive parts will be live at the same potential, hence EQUIPOTENTIAL bonding. In order to achieve this, it is recommended that the resistance of the bonding conductors does not exceed $0.05\,\Omega$.

Table 54H of the *On-Site Guide* and Regulation 547-02-02 in BS 7671 cover the requirements of equipotential bonding. Table 10A of the *On-Site Guide* is also useful. Maximum lengths of copper bonding conductors before $0.05\,\Omega$ is exceeded.

Size mm^2	Length in metres
10	27
16	43
25	68
35	95

The test is carried out with a **Low Resistance Ohm meter** and often can only be carried out on the initial verification; this is because one end of the bonding conductor must be disconnected to avoid parallel paths. When disconnecting a bonding conductor, it is important that the installation is isolated from the supply. On larger installations it is often impossible to isolate the installation and, therefore, the conductor must remain in place. The instrument should be set on the lowest value of Ω possible.

STEP 1

Isolate supply (as safe isolation procedure)

STEP 2

Disconnect one end of the conductor
(*if possible, disconnect the conductor at the
consumers unit, and test from the disconnected end
and the metalwork close to the bonding conductor.
This will test the integrity of the bonding clamp*).

Test leads

STEP 3

Measure the resistance of test leads or null
leads (*these may be long as the only way that we
can measure a bonding conductor is from end to
end*).

Nulled leads

STEP 4

Connect one test lead to the disconnected conductor at the consumer's unit.

Note: Safety notice removed for clarity.

Isolated and locked off   Disconnected   Test lead
conductor

STEP 5

Connect the other end of the test lead to the metalwork that has been bonded (connecting the lead to the metalwork and not the bonding clamp will prove the integrity of the clamp).

Bended metalwork   Other end of test lead

STEP 6

If the instrument is not nulled remember to subtract the resistance of the test leads from the total resistance. This will give you the resistance of the bonding conductor. If the meter you are using has been nulled, the reading shown will be the resistance of the conductor.

Very low value—less than 0.05A

STEP 7

Ensure that the bonding conductor is reconnected on completion of the test.

Whilst carrying out this test a visual inspection can be made to ensure that the correct type of BS 951 earth clamp, complete with label is present, and that the bonding conductor has not been cut if it is bonding more than one service.

If the installation cannot be isolated on a periodic inspection and test, it is still a good idea to carry out the test; the resistance should be a maximum of 0.05Ω as any parallel paths will make the resistance lower. If the resistance is greater than 0.05Ω the bonding should be reported as unsatisfactory and requires improvement.

In some instances the equipotential bonding conductor will be visible for its entire length; if this is the case, a visual inspection would be acceptable, although consideration must be given to its length.

For recording purposes on inspection and test certificates no value is required but verification of its size and suitability is.

Items to be bonded would include any incoming services, such as: water main, gas main, oil supply pipe, LPG supply pipe. Also included would be structural steel work, central heating system, air conditioning, and lightning conductors within an installation (*before bonding a lightning conductor it is advisable to seek advice from a specialist*).

This is not a concise list and consideration should be given to bonding any metalwork that could introduce a potential within a building.

Correct

Incorrect

Continuity of supplementary bonding

There are two general reasons for carrying out supplementary equipotential bonding.

Supplementary bonding 1

This is required when there is an increased risk of electric shock (Regulation 471-08-01).

BS 7671 states that supplementary bonding <u>must</u> be installed in bathrooms/shower rooms and swimming pools.

It should be remembered that, although Regulation 601-04-01 requires that we must bond the exposed and extraneous conductive parts in bathrooms, this Regulation applies only to zones 1, 2, and 3.

Due consideration must be given to other areas where there is an increased risk of electric shock. There is no specific requirement to carry out supplementary bonding in Kitchens. However, if it is thought by the installer that there is an increased risk of electric shock, there is no reason why bonding could not be carried out, it will do no harm providing it is carried out correctly.

On occasions it is often useful to carry out supplementary bonding, particularly under Kitchen sinks. This may not be for electrical reasons, more for visual purposes – bonding is not well understood by many people. A possible scenario might be where you may have travelled 20/30 miles to fit a Kitchen and completed everything to comply with the required regulations. A few days later, before you have been paid for the work you receive a phone call from your customer, informing you that his next door neighbour has spotted that you have not bonded the sink. Of course your customer will believe his neighbour is right and that you have forgotten something or tried to save a bit of money. The choice is now yours, do you try and convince your customer that his neighbour is wrong or do you travel back to the job to carry out the bonding to ensure payment? Perhaps for the sake of a couple of earth clamps and a short length of $4\,\text{mm}^2$, it would have been cheaper just to bond it in the first place.

Where complimentary supplementary bonding is used in this instance a test must be carried out to ensure that the resistance between exposed and extraneous conductive parts is in place and has a resistance of less than $0.05\,\Omega$. The instrument to be used is a low resistance ohm meter.

Probe on solid metal part of tap

Probe on unpainted metal work

Bonding conductor

A visual check must be made to ensure that the correct earth clamps have been used and that they have the correct labels attached.

It is perfectly acceptable to use the pipe work and structural steelwork within the area as a bonding conductor, and bonding can be carried out adjacent to the area providing that the integrity of the pipe work/steelwork can be assured. An airing cupboard would be a good example of a suitable place to bond.

If it is necessary for the lighting point or electric shower in a bathroom (*remember if they are not within zones 1, 2, or 3 no bonding is required*), there is no reason why the bonding conductor could not be simply attached to a pipe within the roof space, which is bonded elsewhere and passes near the item which requires bonding. The correct bonding clamps and labels should always be used.

If using pipe work of plumbing and heating systems as bonding, continuity of the pipe work must be verified. The resistance values are the same as if copper cables were used. Tests must be made between exposed and extraneous conductive parts to ensure that the resistance does not exceed $0.05\,\Omega$. This is a simple test carried out using a low resistance ohm meter. A probe of one lead should be placed on one metal part and the probe of the other lead placed on an adjacent metal part. The resistance must be no greater than $0.05\,\Omega$.

Problems can arise if the pipe work is altered and plastic push fittings are used. Clearly these will not conduct and the bonding continuity could be compromised. If ever a plastic plumbing fitting is used on copper pipe work, consideration should be given to the installation of a bonding conductor installed across the fitting.

It is a common belief that water in pipe work will conduct; in fact, the current that would flow through water in a 15 mm diameter pipe which has a plastic joint in it is very small.

To find out just how much, I set up a simple controlled experiment, two short lengths of 15 mm pipe were joined using a 15 mm plastic push fit coupler. The pipe was then filled with water and the two ends of the joined pipe were connected to a 230 volt supply. The current flowing was measured to be 0.003 Amperes (3 mA). The current flow would increase if the water had central heating additives in it, but not considerably.

In bathrooms/shower rooms where the plumbing has been carried out using plastic pipe, the pipe work does not need supplementary bonding; however, it should be remembered that electrical appliances and any extraneous/exposed conductive parts within zones 1, 2 and 3 must still be bonded. Figure 4d and 4e in section 4 of the *On-Site Guide* are good places of reference for this.

For the sizing of supplementary bonding Table 10b in Appendix 10 of the *On-Site Guide* should be used. As a general rule, supplementary bonding within a bathroom should be 2.5 mm^2 if mechanically protected or 4 mm^2 if not. It is usually easier to use 4 mm^2 to save the trouble of mechanically protecting the bonding conductor (see Regulation 547-03 of BS 7671).

Supplementary bonding 2

Used where, due to various circumstances, the required disconnection time cannot be met by conventional methods, and where it is not desirable to use a residual current device. Supplementary equipotential bonding may be used as a means of compliance (Regulation 413-02-04 (i)).

Wherever supplementary bonding is used in this instance it must be tested to ensure that the resistance between two parts is less than the value R obtained by the following formulae:

$$R = \frac{50}{I_a}$$

R = maximum resistance of bonding conductor; 50 = the safe touch voltage (*this may be 25 volts in special locations*); I_a = fault current needed to operate the protective device within 5 seconds (Regulation 413-02-28).

If the circuit is protected by a Residual current device I_a may be substituted by $I\Delta_n$.

The fault current can be found in Appendix 3 in BS 7671 or by using the calculation:

$$I_a = \frac{U_{OC}}{Z_S}$$

U_{OC} = open circuit voltage of supply transformer; Z_S = maximum earth loop impedance for the protective device.

The test should be carried out between exposed and extraneous conductive parts, using a low resistance ohm meter with long leads to find out the resistance between them. If the resistance is higher than that required, then supplementary bonding is required.

Exposed conductive parts

Extraneous conductive parts

Screw fixings are earthed

Unpainted metal work

Example 1

A circuit on a TN system is to be altered. The current carrying capacity and the volt drop of the cable are adequate for the load. However, the Z_S value is too high for the 20A BS 1361 protective device which is being used to protect the circuit.

The maximum resistance permissible between exposed and extraneous conductive parts must be calculated.

The first step is to find the current that would cause automatic disconnection of the supply.

$$I_a = \frac{U_{OC}}{Z_S}$$

$$I_a = \frac{240}{2.93} = 81.91 \, amps$$

This value can also be found in Figure 3.1 of Appendix 3 of BS 7671 (it is rounded up to 82 amps).

The maximum permissible resistance between conductive parts can now be found by:

$$R = \frac{50}{I_a}$$

$$R = \frac{50}{82} = 0.6\Omega$$

Supplementary bonding is installed where there is a risk of simultaneous contact with any extraneous and exposed conductive parts. Its purpose is to ensure that the potential between any of these parts does not rise above a safe value. In most cases, this value is 50 volts, although some chapters in Part 6 of BS 7671 require a maximum potential of only 25 volts.

Determining if a part is extraneous, or just a piece of metal

A test should be made using an insulation resistance tester set on mΩ, supplying 500 volts.

Connect one test lead to the metal part and the other lead to a known earth. If the resistance value is 0.02 mΩ (20,000 Ω) or greater, no supplementary bonding is required. If less than 0.02 mΩ, supplementary bonding should be carried out.

If we use Ohm's law we can see how this works:

$$\frac{V}{R} = I: \qquad \frac{500}{20,000} = 0.025 \, A$$

This shows that a current of 25 mA would flow between the conductive parts; this would of course only be 0.012 amp if the fault was on a single phase 230 volt supply. This current is unlikely to give a fatal electrical shock.

The test must not be confused with a continuity test. It is important that an insulation resistance tester is used.

Continuity of circuit protective conductors

This test is carried out to ensure that the CPC of radial circuits are intact and connected throughout the circuit. The instrument used for this test is a **low resistance ohm meter** set on the lowest value possible.

This is a dead test and must be carried out on an isolated circuit.

Testing can be carried out using two methods.

Method one

STEP 1

Using a short lead with a crocodile clip on each end, bridge phase and CPC together at one end of the circuit (*it does not matter which end, although it is often easier to connect at the distribution board as this will certainly be one end of the circuit*).

The resistance of this lead plus the resistance of the test leads should be subtracted, or the instrument nulled before the $R_1 + R_2$ reading is recorded.

Connect to main earth terminal Connect to phase conductor

STEP 2

At **each** point on the circuit test between phase and CPC.

Terminal of switch return   Earthing terminal

Earthing terminal   Switch return

Keep a note of the readings as you carry out the test, they should increase as you move further from the connected ends. The highest reading obtained should be at the furthest end of the circuit and will be the $R_1 + R_2$ of the circuit. This value should be recorded on the schedule of test results. If the highest reading is obtained at a point which is not the furthest from the circuit, further investigation should be carried out as it may indicate a loose connection (high resistance joint).

In some instances only the value of R_2 may be required. Where the phase conductor is the same size as the CPC the total measured resistance can be divided by 2 as the phase and CPC resistance will be the same. If the phase and CPC are of different sizes (this is usual in twin and earth thermoplastic cable) the R_2 value can be calculated using the following formula:

$$R_2 = R_1 + R_2 \times \frac{A\ phase}{A\ CPC + A\ phase}$$

R_2 = resistance of CPC in ohms; $R_1 + R_2$ = measured value of resistance in ohms; $A\ phase$ = Area of phase conductor in mm^2; $A\ CPC$ = Area of CPC mm^2.

Example 2

A radial circuit is wired in 2.5 mm^2 phase and 1.5 mm^2 CPC. The test resistance of $R_1 + R_2$ is 0.37 Ω.

To calculate the resistance of the CPC on its own:

$$R_2 = 0.37 \times \frac{2.5}{2.5 + 1.5}$$
$$R_2 = 0.37 \times 0.625 = 0.23\Omega$$

If the CPC is smaller than the phase conductor, the resistance of the CPC conductor will always be greater than the phase conductor as it has a smaller cross-sectional area.

Another method of determining R_2 is described in the ring circuit test.

Method two

This method will prove CPC continuity and is usually only used where the circuit is wired in steel enclosures where parallel paths to the CPC may be present and the $R_1 + R_2$ value would not be a true value. Or where the CPC resistance is required for use with Table 41C of the Wiring Regulations.

This method uses a long lead. One end is connected to the earth terminal of the distribution board, and the other connected to a **low resistance ohm meter**. The short lead of the ohm meter is then touched onto each fitting to ensure that it is connected to the CPC. The highest reading minus the resistance of the leads can be recorded as the R_2 reading.

If the furthest point of the circuit is known, and no parallel paths exist, the $R_1 + R_2$ reading can be carried out first using **Method one**, and then a test between earthed metal at each point can be made to ensure that the CPC is connected to each point on the circuit, using **Method two**. This method is particularly useful where there are a lot of enclosed metal fittings and dismantling them would be impractical.

Ring final circuit test

The purpose of this test is to ensure that:

- The cables form a complete ring.
- There are no interconnections.
- The polarity is correct on all socket outlets.

When this test is carried out correctly it also gives you the R_1 and R_2 value of the ring and identifies spurs.

Table 8A in the *On-Site Guide* provides information on final circuits for socket outlets. This table states that a ring circuit is to be wired in 2.5 mm^2 phase conductor and 1.5 mm^2 CPC as a minimum size. This type of circuit is an A1 ring and should be protected by a 30/32 amp overcurrent protective device.

Complete ring circuit

A test must be carried out on the conductors to verify that they form a complete loop. If it is found that they do not, overloading of the cables could occur. In installations where more than one ring circuit has been installed, it is possible for the ends of the ring to become muddled, resulting in the circuits being supplied through two protective devices.

Complete ring circuit

The whole point of a ring circuit is that it can be wired in small CSA cables but carry a reasonably high current, this is because we have two 2.5 mm^2 cables wired in parallel (*Regulation 473-01-06*). If we look at Table 4D5A in BS 7671, the value of current that 2.5 mm^2 cable can carry is 20 amps in the worst type of conditions.

If we use two of these conductors in parallel, we will have a total current carrying capacity of 40 amps. As one of the jobs of the protective device is to protect the cable, this situation will be fine because the protective device is smaller than the total current carrying capacity of the cables in parallel.

Broken ring circuit

If, however, we found the ring to be broken, the protective device could not do its job as it is rated at 32amps and the cable is rated only at 20amps. Hence overloading!

Interconnections

Occasionally a situation will be found where there is a ring within a ring, in other words the ring is interconnected.

Broken ring circuit

Interconnected ring circuit

This situation, as it is, will not present a danger. However, it will make it very difficult for a ring final circuit test to be carried out as, even if the correct ends of the ring are connected together, different values will be found at various points of the ring. If one loop is broken, a test at the consumer's unit will still show a complete ring. It will not be until further tests are performed that the interconnection/broken loop will be found.

Polarity

Each socket outlet must be checked to ensure that the conductors are connected into the correct terminals. Clearly if they are not, serious danger could occur when appliances are plugged in.

It could be that phase and neutral are the wrong polarity; the result of this is that the neutral would be switched in any piece of equipment with a single pole operating switch.

If the live conductors and CPC are connected with reverse polarity, then the case of any Class 1 equipment could become live and result in a fatal electric shock.

Performing the test

The instrument required is a low resistance ohm meter set on the lowest scale, typically $20\,\Omega$. Be sure to zero the instrument or subtract the resistance of the leads each time you take a reading.

> This is a dead test! Safe isolation must be carried out before working on this circuit

STEP 1

Isolate circuit to be tested.

STEP 2

Identify legs of ring.

STEP 3

Test between ends of phase conductor and note the resistance value.

Instrument set to Ω for whole test.

Ends of phase conductor

STEP 4

Test between ends of neutral conductor. This value should be the same as the phase conductor resistance as the conductor must be the same size (*see Note 1*).

Ends of neutral conductor

STEP 5

Test between the ends of the CPCs. If the conductor size is smaller than the live conductors (*as is usually the case when using twin and earth cable*), the resistance value will be higher (*see Note 2*); make a note of this reading.

Ends of CPCs

STEP 6

Join P of leg 1 to N of leg 2.

Test between N of leg 1 and P of leg 2. The measured resistance should be double that of the phase conductor.

Resistance double that of phase conductor   P2   P1 joined to N2   N1

STEP 7

Join N of leg 1 to P of leg 2 together (*leaving N2 and P1 joined*).

Test between joined ends.

The measured value should be $\frac{1}{4}$ of test between N of leg 1 and P of leg 2.

| P2 joined to N1 | Resistance 1/4 of that tested between N1 and P2 | Test between joined ends | P1 still joined to N2 |

STEP 8

Leave the ends joined.

Test between P and N at each socket outlet, the resistance should be the same at each socket (*see Note 1*).

A higher reading should be investigated, although it will probably be a spur it should be checked as it may be a loose connection (*high resistance joint*).

| Resistance value the same at each socket | N | P | Test at each socket outlet |

STEP 9

Disconnect the ends and repeat the test using phase and CPC conductors (*see Note 3*).

Phase   CPC joined to phase   CPC

P2 joined   P1 joined
to CPC 1   to CPC 2

CPC   P   Test at each
             socket outlet

The highest value (*which will be the spur*) will be the R_1 and R_2 value for this circuit.

Notes

1. If with ends connected (P1/N2 and P2/N1) a substantially different resistance value is measured at each socket outlet, check that the correct ends of ring are connected. A difference of 0.05 Ω higher or lower would be acceptable.
2. In a twin and earth cable the CPC will usually have a resistance of 1.67 times that of the phase conductor as it has a smaller cross-sectional area.
3. When phase and CPC conductors are not the same size a higher resistance value will be measured between Phase and CPC than Phase and neutral. It will also alter **slightly** as the measurement is taken around the ring, the resistance will be lower nearer the joined ends and will increase towards the centre of the ring. The centre socket of the ring will have the same resistance value as the test between the joined ends.
4. If the circuit is contained in steel conduit or trunking parallel paths may be present, this would result in much lower $R_1 + R_2$ resistance values.
5. Some certificates may require r_n to be documented. This is the resistance of the neutral loop measured from end to end.

Example 3

Let's use a 2.5/1.5 mm^2 twin and earth cable 22 metres long. If we look at Table 9A in the *On-Site Guide* we will see that the resistance of a copper 2.5 mm^2 conductor has a resistance of 7.41 mΩ per metre.

The resistance of the phase conductor will be:

$$\frac{7.41 \times 22}{1000} = 0.163\Omega$$

Divide the largest conductor by the smallest to find the ratio of the conductors (how much bigger is the larger conductor?).

$$\frac{2.5}{1.5} = 1.67$$

The 2.5 mm^2 conductor is 1.67 $\times$ larger than the 1.5 mm^2 conductor; therefore, it must have 1.67 $\times$ less resistance than the 1.5 mm^2 conductor.

If we now multiply the resistance of the phase conductor by 1.67:
$0.163 \times 1.67 = 0.27\,\Omega$ *this is the resistance of the 1.5 mm² conductor.*

We can check this by looking at Table 9A of the On-Site Guide once again, and we can see that the resistance of 1.5 mm² copper is 12.10 mΩ per metre. Therefore, 22 metres of 1.5 mm² copper will be:

$$\frac{22 \times 12.10}{1000} = 0.266\,\Omega$$

As a final check, if we look at Table 9A of the On-Site Guide for the resistance of a 2.5 mm²/1.5 mm² cable, we will see that it has a resistance of 19.51 mΩ per metre, and that 22 metres of it will have a resistance of:

$$\frac{22 \times 19.51}{1000} = 0.429\,\Omega$$

The resistance value of the 2.5 mm² is 0.163 Ω; and the resistance value of the 1.5 mm² is 0.266 Ω.

If we add them together: 0.163 Ω + 0.266 Ω = 0.429 Ω. Finally, 0.429 Ω is the resistance of our 2.5 mm²/1.5 mm² measured as one cable.

Insulation resistance test

This is a test that can be carried out on a complete installation or a single circuit, whichever is suitable or required. The test is necessary to find out if there is likely to be any leakage of current through the insulated parts of the installation. A leakage could occur for various reasons.

A good way to think of this test is to relate it to a pressure test – we know that voltage is the pressure where the current is located in a cable. On a low voltage circuit, the expected voltage would be around 230 V a.c. The voltage used in an insulation test on a 230 V circuit is 500 V, which is more than double the normal circuit voltage. Therefore, it can be seen as a pressure test similar to a plumber pressure testing the central heating pipes.

Low insulation resistance

Cable insulation could deteriorate through age. A low insulation resistance caused through age will often be found in installations where rubber-insulated cables have been used. Cables which are crushed under floor boards, clipped on edge, or worn thin where pulled through holes in joists next to other cables, can give a very low reading.

Low insulation resistance could be found if a building has been unused for a period of time, due to the installation being affected by dampness in the accessories. Low insulation resistance readings will also often be found where a building has been recently plastered. In theory, long lengths of cables or circuits in parallel could give low readings due to the amount of insulation (*the longer the circuit or the more circuits, the more insulation there will be for leakage to occur*).

The instrument used to carry out this test is an **insulation resistance tester**. To comply with the requirements of the Health and Safety Executive the instrument must be capable of delivering a current of 1 mA when a voltage of 500 V.d.c is applied to a resistance of 0.5 MΩ. Table 71A in BS 7671 gives the test voltages and minimum acceptable resistance values.

The values are shown here:

	Circuits between 0 V and 50 V a.c.	Circuits between 50 V a.c. and 500 V a.c.	Circuits between 500 V and 1000 V a.c.
Required test voltage	250 V d.c.	500 V d.c.	1000 V d.c.
Minimum acceptable	0.25 MΩ	0.5 MΩ	1 MΩ
In 17th Edition of the Wiring Regulations	(0.5 MΩ)	(1 MΩ)	(2 MΩ)

Domestic installations

Remember that testing should be carried out from the day the installation commences (Regulation 711-01-01).

Testing a whole installation

In <u>new</u> domestic installations it is often easier to carry out insulation resistance testing on the whole of the installation from the meter tails <u>before</u> they are connected to the supply. If this is the preferred choice the test should be carried out as follows:

- Safe isolation must be carried out before commencing this test.
- Inform any occupants of the building that testing is to be carried out.
- Ensure that all protective devices are in place and switched on.
- Remove all lamps from fittings where accessible.
- If the lamps are not accessible or if a luminaire with control gear (fluorescent) is connected, open the switch controlling the luminaire (*Note 6*).

The same applies to extra low voltage transformers.

- Where dimmer switches are fitted it is important that they are either removed and the switch wires joined, or that the switch is bypassed (*Note 7*).
- Any accessories with indicator lamps are switched off (*Note 8*).
- Passive infra red detectors (PIRs) are removed or bypassed (*Note 9*).
- All fixed equipment such as cookers, immersion heaters, boilers and television amplifiers are isolated.
- Shaver sockets are disconnected or isolated (*Note 9*).
- Items of portable equipment are unplugged.

Great care must be taken as, during this test, 500 V will be passed through any electrical equipment which is left connected. This could damage the equipment or, at the very least, cause low readings to be obtained during the test. Once all precautions have been taken proceed with the test as follows:

Notes

1. The control equipment within discharge lamps will cause very low readings. It is quite acceptable to isolate the fitting by turning off the switch. This is more desirable than disconnecting the fitting. After the test between live conductors is completed the control switch for the luminaire should be closed before carrying out the test between live conductors and CPC. This is to ensure that all live conductors are tested for insulation resistance to earth.
2. Most dimmer switches have electronic components in them and these could be damaged if 500 V were to be applied to them. It is important that wherever possible the dimmer switches are removed and the phase and switch return are joined together for the test.
3. Neon indicator lamps will be recognized as a load by the test instrument and will give a very low insulation value. All that is required is for the switch on the accessory to be turned off.
4. Passive infra red detectors will give very low readings and may be damaged by the test voltage. Either disconnect it or test between live conductors and earth only on circuits containing PIRs. The same applies to shaver sockets.

STEP 1   Set insulation resistance tester to 500 V.

Some instruments have settings for Meg ohms and some
are self ranging. If yours requires setting then 200 MΩ
or higher is the setting to use.

STEP 2

To ensure that the test results are accurate it is
important to ensure correct operation of instrument
and the integrity of the leads. Push the test button with
the leads disconnected. The resistance shown on screen
should be the highest that the instrument can measure.

Over range   Leads not connected

STEP 3

Join leads and operate instrument again, the resistance
shown on screen should be the lowest value possible
(0.0 mΩ) in all cases.

Closed circuit   Leads connected

STEP 4

When testing the whole installation from the disconnected tails, it is important that the main switch is in the on position and that the protective devices are in place. If they are circuit breakers they must be in the 'on' position.

Test between live conductors (tails) and operate any two-way and intermediate switching. This is to ensure that all switch wires and strappers are tested and that the switch returns have been correctly identified and connected (*no neutrals in the switches*).

Connected to live conductors   Main switch and circuits on

STEP 5

Join live conductors (tails) together, connect IR tester leads, one on live conductors and the other on the earthing conductor, carry out the test and again operate all two-way and intermediate switching.

Earthing   Live conductors   Main switch
conductor   joined together   and circuits on

Table 71A of BS 7671 give the acceptable insulation resistance as 0.5 MΩ. This is for a single circuit or a complete installation. *Guidance Note 3* in BS 7671 recommends that any circuit under 2 MΩ should be investigated, as such a low insulation resistance value could indicate a latent defect. Although these values will comply with the requirements of BS 7671, they are very low values for the majority of domestic circuits and installations.

Depending on the type of circuit it may be wise to carry out further investigation on the circuit with the lower reading. It would be very rare that an insulation resistance value of between 0.5 MΩ and 2 MΩ would be acceptable. There would have to be a very good reason for this, and it would need to be monitored to ensure that there was not continued deterioration.

A low value would possibly be acceptable if, for instance, a building had been empty and unused for a period of time, or perhaps an underground cable which had been unused for some time was to be reconnected to supply an outbuilding, or any other outdoor circuit. It could be that, after a period of continued use, the insulation resistance rises; only regular testing would show this.

Example 4

An installation consisting of five circuits is tested as a whole, and the insulation resistance value between live conductors and earth is found to be less than 2 MΩ.

The installation is split into individual circuits and each circuit is tested again, all circuits are found to be greater than 200 MΩ except a mineral insulated cable serving electrically opening gates 30 metres from the building. This circuit is found to have an insulation resistance value of 2 MΩ. In this instance, the value would be compared to previous results to see if any deterioration had occurred since the last test. In the likely event that previous results are not available, the result could be recorded for future reference.

If, however, the low value circuit was a new circuit wired in PVC thermoplastic cable, 6 metres long further investigation should be carried out as the insulation resistance for this circuit should be beyond the scale of the test instrument; in this case > 200 MΩ.

When testing is carried out on any installations it is important that the results are thought about and not just recorded. As shown in the example above, an element of careful thought and sensible judgement has to be included in the testing process.

If a high insulation resistance value for a complete installation is measured, it is permissible to enter this value for all circuits on the schedule of test results.

Testing of individual circuits

If it is necessary to test individual circuits, the same process can be applied to new and existing circuits; and the same safety precautions must be taken.

Ensure that safe isolation of the circuit to be tested is carried out. It is preferable for safety reasons to isolate the whole of the distribution board if possible. If not, each circuit can be isolated individually by removing the fuse or turning off and locking off the circuit breaker.

It will be necessary to disconnect the neutral _or_ CPC of the circuit to be tested. This is because the neutrals for all of the circuits will be connected to a common neutral bar and the CPCs for all circuits will be connected to a common earth bar.

Check that equipment vulnerable to testing, or any equipment that could produce a low insulation resistance reading is disconnected or isolated.

Carry out a test between live conductors. And then live conductors and earth.

> If there was an insulation fault between any neutral and CPC in the system it would show up as a fault on all circuits, disconnection of the neutral or CPC will stop this happening.

Neutral   Between live conductors   Phase

Neutral   Between live conductor and earth   Earthing terminal

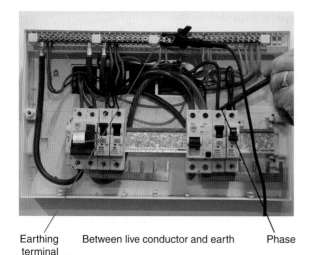

Earthing terminal   Between live conductor and earth   Phase

Earthing terminal   Live conductors

Where testing between live conductors and earth the live conductors can be joined and then tested to earth. Alternately, they can be tested separately, whichever is the easiest.

If, for some reason, there is a piece of equipment connected to the system that cannot be isolated from the circuit under test, do not carry out the test between live conductors – only test between live conductors and earth. This is to avoid poor readings and possible damage to equipment. This test should only be carried out on individual circuits, not whole installations, as it is important to test as much of the installation as possible.

Testing of 3 phase installations

When testing 3 phase and neutral (TP&N) installations or circuits the same safety precautions apply as when testing single phase installations or circuits.

The test for the whole installation can be carried out on the isolated side of the main switch. If this is the chosen method it is important to ensure that all of the protective devices are in the 'on' position. Safe isolation must be carried out before commencing this test.

STEP 1

The instrument should be set on 500 volts DC.

Phase L2 and L3   Locked off and isolated

STEP 2   Test between all phase conductors.

Phase L1 and L3   Locked off and isolated

Phase L1 and L2   Locked off and isolated

STEP 3

Test between all phase conductors and neutral, the phase conductors can be joined together for this test to save time.

Locked off   All phases   Neutral
and isolated   joined together

STEP 4

Test between all phase conductors and earth. The phase and neutral conductors can be joined to save time.

Earth   Locked off   All phases joined
         and isolated together

STEP 5

Test between neutral and earth.

Earth All phases   Locked off   Neutral
     joined together and isolated

STEP 6 The resistance values should be $2\,M\Omega$ or greater for each circuit and the whole installation must have an insulation resistance of greater than $0.5\,M\Omega$.

STEP 7 If a circuit is found to have an insulation resistance of less than $2\,M\Omega$ it should be investigated.

STEP 8 Ensure that any links used for testing are removed prior to the switching on of the supply.

Be aware that if the live conductors are joined and then tested to earth, theoretically the insulation resistance value may be lower due to the conductors being in parallel.

Example 5

A three-phase sub-main is tested and the results are as follows:

L1 to earth is $130\,M\Omega$
L2 to earth is $80\,M\Omega$
L3 to earth is $50\,M\Omega$
N to earth is $100\,M\Omega$

If these conductors were now joined and tested to earth the value would be as given below.

Calculation is:

$$\frac{1}{R1} + \frac{1}{R2} + \frac{1}{R3} + \frac{1}{R4} = \frac{1}{Rt}$$

Put in the values:

$$\frac{1}{130} + \frac{1}{80} + \frac{1}{50} + \frac{1}{100} + \frac{1}{0.05} = 19.92\,M\Omega$$

Enter it this way into a calculator:

$$130X^{-1} + 80X^{-1} + 50X^{-1} = X^{-1}\ Answer\ is\ 19.92\,M\Omega$$

This value is still acceptable but lower because the conductors are in parallel.

When completing a Schedule of Test Results, it is important that you document exactly which insulation tests have been carried out. If it is not possible to carry out a test between any conductors a note of this should be made in the remarks column of the schedule.

Testing of site applied insulation

This type of testing is only required where insulation is applied during the erection of equipment on site. It is not required to be carried out on site built assemblies where pre-tested equipment is built on site.

The test is carried out by wrapping metallic foil around all surfaces which has been insulated. Then joining all live conductors together and applying a voltage of 3750 V a.c. between the foil and the conductors for a duration of 60 seconds. The insulation is satisfactory if a breakdown of the insulation does not occur within this time. The instrument used to carry out this test is an applied voltage tester.

Further information on this type of test can be found in the IEE *Guidance Note 3* in BS 7671).

Polarity tests

This is a test carried out to ensure that:

- Protective devices are connected to the phase conductors of the circuits which they are protecting.
- Switches in circuits are in the phase conductor.
- ES lampholders have the centre pin connected to the phase conductor (except for E14 and E27 lampholders complying with BSEN 60238) (Regulation 553-03-04).
- Accessories such as fused connection units, cooker outlets and the like are correctly connected.

Remember! If you are working on a consumer unit or a distribution board, the whole board should be isolated.

In many cases this test can be carried out at the same time as the CPC continuity test. The instrument required is a low resistance ohm meter.

Ensure that the circuit that is to be tested has been isolated.

Polarity test on a radial circuit such as a cooker or immersion heater circuit

STEP 1

At the origin of the circuit, link the phase and CPC using a short lead with crocodile clips at each end. Whilst connecting the lead visually check that the phase conductor is colour identified and connected into the protective device.

STEP 2

At the furthest point of the circuit remove the cooker or immersion heater switch, visually check that the phase conductor is identified and connected into the correct terminal.

Remember to null the leads of the instrument if the value is to be used for $R_1 + R_2$.

Earthing terminal — linked by lead — Phase

STEP 3

Connect the leads of the test instrument to the incoming phase and CPC terminals at the switch. Test to ensure continuity. The instrument should show a very low reading. This will also be the $R_1 + R_2$ for the circuit.

STEP 4

Remove the lead at the consumer unit and test again. The circuit this time should be open and the instrument reading will be high. This will prove that the correct circuit is being tested.

Test instrument probe at incoming CPC terminal of switch

Test instrument probe at incoming phase terminal of switch

Polarity test on a lighting circuit

STEP 1

At the origin of the circuit connect the phase and CPC, this can be done with a short lead with crocodile clips at each end.

Phase   Linked by lead   CPC

STEP 2

At the ceiling rose or light fitting place the probes of the instrument on to the earthing terminal and the switched live.

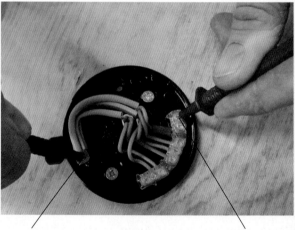

Test instrument probe at earthing terminal   Connection to light fitting place   Test instrument probe at switched live

STEP 3

Close the switch controlling the light and the instrument should read a very low resistance (this will also be the R_1 and R_2 reading for the circuit). When the switch is opened the instrument reading should be very high.

This test can also be carried out at the switch if required:

STEP 1

Place a link between the phase and CPC of the circuit.

Earthing terminal ———— linked by lead ————— Phase

STEP 2

Place the probes of the test instrument on the earth terminal at the switch and the switch return terminal.

Test instrument probe at switch return terminal   Connection to switch   Test instrument probe at earth terminal

STEP 3

Close the switch and a low resistance reading should be shown on the instrument.

Low resistance
reading

STEP 4

Open the switch and the instrument reading will show over range as the circuit should be open circuit.

Over range

Live polarity test

This test is usually carried out at the origin of the installation before it is energized to ensure that the supply is being delivered to the installation at the correct polarity.

The instrument to be used is an approved voltage indicator or test lamp that complies with HSE document GS 38. It is acceptable for an earth loop impedance meter to be used as these instruments also show polarity.

> Great care must be taken whilst carrying out this test as it is a live test.

STEP 1

Place the probes of the voltage indicator onto the phase and neutral terminal of the incoming supply at the main switch. The device should indicate a live supply.

Neutral on main switch   Phase on main switch   Live supply showing

STEP 2

Place the probes of the voltage indicator onto the phase and earth terminal of incoming supply at the main switch. The device should indicate a live supply.

Phase on main switch   Live supply showing   Earthing terminal

STEP 3

Place the probes of the voltage indicator onto the earthing terminal and the neutral terminal at the main switch. The device should indicate no supply.

Neutral
main switch

No supply
showing

Earth
terminal

Earth electrode testing

Earth fault loop impedance tester

For many installations the resistance of the earth electrode can be measured using an earth fault loop impedance test instrument. It is perfectly acceptable to use this type of instrument on a TT system where reasonably high resistance values could be expected.

The test is performed in exactly the same way as the external earth fault loop Z_e test.

STEP 1 Isolate the installation.

STEP 2 Ensure that the earthing conductor is correctly terminated at the earth electrode.

STEP 3 Disconnect the earthing conductor from the main earthing terminal.

STEP 4 Connect a lead of the earth fault loop meter to the disconnected earthing terminal.

STEP 5 Place the probe of the other lead on to the incoming phase conductor at the supply side of the main switch and carry out the test.

STEP 6 Record the result.

STEP 7 Reconnect the earthing conductor and leave the installation in safe working condition.

If a three lead test instrument is used, read the instrument instructions before carrying out the test. It may be that the leads must be connected on to the phase, neutral and earthing terminals, or possibly the neutral and earth lead of the test instrument, should be joined together.

Regulation 413-02-16 from BS 7671 tells us that the maximum permissible resistance value of the earth electrode must be no greater than 50/IΔn. This is to ensure that any exposed metalwork does not rise to a potential of greater than 50 volts.

    50 is the maximum voltage
    IΔn is the trip rating of the residual current device
    Z_S is the earth fault loop impedance

If the rating of the device is 100 mA the calculation is:

$$\frac{50}{0.1} = 500 \ \Omega$$

The maximum permissible value to comply with the regulations using this calculation is 500 Ω. Although this value would be deemed acceptable, it may not be reliable as it could rise to an unacceptable value if the soil dries out.

An acceptable value for RCDs up to 100 mA is stated as a maximum of 200 Ω. If the resistance is above this it should not be accepted under most circumstances. The installation of an additional or larger electrode may bring the resistance value down to an acceptable value.

The maximum calculated values for earth electrodes are:

Operating current of the RCD	Electrode resistance in ohms
100 mA	500
300 mA	160
500 mA	100

For special locations where the maximum touch voltage is 25 V the electrode resistance should be halved. Electrode tests should be carried out in the worst possible conditions. The worst condition for an earth electrode is when the soil is **dry**. Where lower values of earth electrode resistances are required, an earth electrode tester should be used.

Measurement using an earth electrode test instrument

This test requires the use of three electrodes: the earthing electrode under test, a current electrode and a potential electrode.

STEP 1 The earthing electrode (E) should be driven into the ground in the position that it is to be used. Attention should be paid to the length of the electrode which is in the ground.

STEP 2 The current electrode (C^2 spike) should be pushed into the ground at a distance of ten times the depth of the electrode under test away from it.

STEP 3 The potential electrode (P^2) should be pushed into the ground midway between E and C^2.

STEP 4 The leads of the test instrument should be connected to the appropriate electrodes.

STEP 5 Measure the value of resistance.

STEP 6 Move P^2 10% closer to C^2.

STEP 7 Measure the value of resistance.

STEP 8 Move P^2 back to 10% closer to E than the mid-point.

STEP 9 Measure the value of the resistance.

A calculation must now be carried out to find the percentage deviation of the resistance values.

Example 6

Three measurements are taken: 0.8 ohms, 0.86 ohms and 0.78 ohms. These must now be added together and an average value calculated.

Total value of three readings = 2.44

Find the average.

$$\frac{2.44}{3} = 0.81$$

The average value is 0.81 ohms.

Now find the highest difference between the average value and the measured values. In this case it will be 0.86 − 0.81 = 0.05.

The percentage of this value to the average value must now be found:

$$\frac{0.05}{0.81} \times 100 = 6.1\%$$

This value is higher than 5% of the average value and it is not advisable to accept, it as a percentage deviation of greater than 5% is deemed to be inaccurate. To overcome this, the distance between the electrode under test (E) and the current spike (C^2) should now be increased and the first three tests repeated to obtain a more accurate reading.

If the required resistance value cannot be obtained by the use of a single electrode, additional electrodes may be added at a distance from the first electrode equal to its depth.

Earth fault loop impedance Z$_e$

This is a live test and great care must be taken

Z_e is a measurement of the external earth fault impedance (resistance) of the installation. In other words, it is the measured resistance of the supply transformer winding, the supply phase conductor, and the earth return path of the supply. It is measured in ohms.

Earth fault path for a TT system

This system uses the mass of earth for the fault return path.

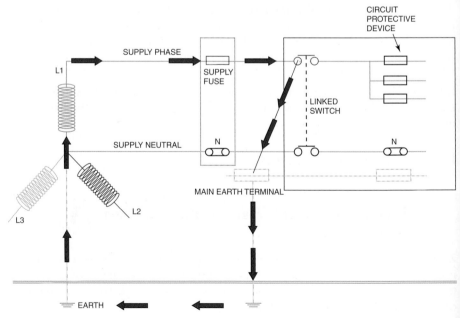

Earth fault path for a TNS system

This system uses the sheath of the supply cable for the earth fault return path.

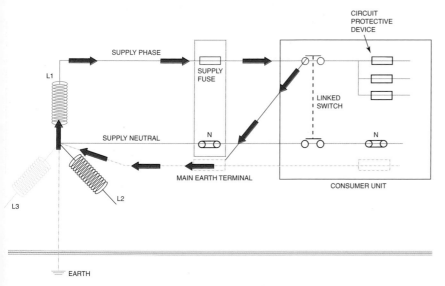

Earth fault path for a TNCS system

This system uses the neutral (PEN) conductor of the supply for its earth fault return path.

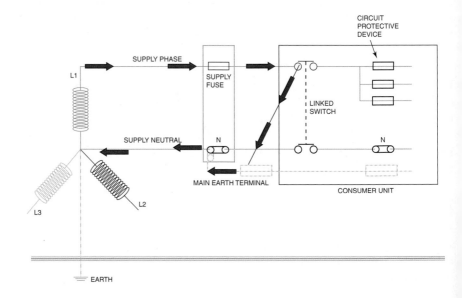

To carry out this test correctly the installation should be isolated from the supply and the main earth disconnected from the main earth terminal (MET). This is to avoid the possibility of parallel paths through any earthed metalwork within the installation.

Often in industrial and commercial installations – where isolation may be impossible due to the building being in use – the only time that this test can be carried out with the main earthing conductor disconnected, is during the initial verification. It is important that the Z_e with the earthing conductor is disconnected during the initial verification as this will give a reference value for the life of the installation. If, during subsequent tests, the earthing conductor cannot be disconnected, a test can still be carried out but the parallel paths should give a lower impedance value. If a higher value is recorded it will indicate a deterioration of the supply earth.

The instrument used for this test is an earth fault loop impedance meter, and it is important that the person using the instrument has read and understood the operating instructions. There are many types of test instruments on the market and they all have their own characteristics.

Some instruments have three leads which must be connected to enable this test to be carried out correctly. Some instruments require that the leads are connected to the phase, neutral and earth of the circuit to be tested. Other instruments require the phase lead to be connected to the phase conductor and the earth and neutral leads to be connected to the earthing conductor of the circuit to be tested (*it is important to read the instructions of the instrument being used*).

If using a two-lead instrument it should be set on Z_e. One lead should be connected to the main earthing conductor and the other to the incoming phase on the supply side of the main switch.

Probes can be used for this providing they meet the requirements of GS 38, i.e.:

- Insulated
- Fused
- Finger guards
- Maximum of 4 mm exposed tips or retractable shrouds
- Long enough to carry out test safely
- Undamaged

Performing the test

This is a live test
and care should be
taken when
carrying it out

STEP 1 Isolate the supply.

STEP 2 Disconnect the earthing conductor.

STEP 3 Set the instrument to loop test.

STEP 4 If you are using a two-lead instrument the leads should be connected as in the figures below. If you are using a three lead instrument, then the leads should be connected as shown here.

Connections for 2-lead instrument   Disconnected earthing conductor

Connections for 3-lead instrument   Leads joined

STEP 5 The measurement obtained is Z_e and can be entered on the test certificates in the appropriate place.

STEP 6 Reconnect earthing conductor.

It is important that the instructions for the test instruments are read and fully understood before carrying out this test.

Care should be taken to reset time clocks, programmers, etc. when the supply is reinstated.

Circuit earth fault loop impedance

This is a live test and great care must be taken

Z_S is the value of the earth loop impedance (resistance) of a final circuit including the supply cable.

The earth fault loop (Z_S) path for a TT system.

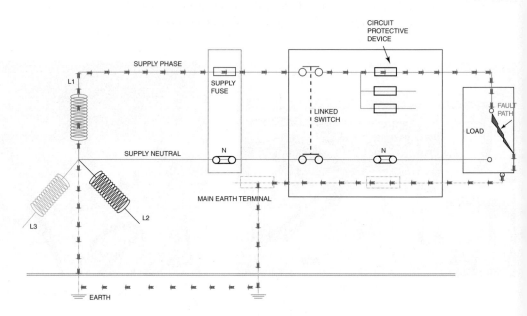

The earth fault loop (Z_S) path for a TNS system.

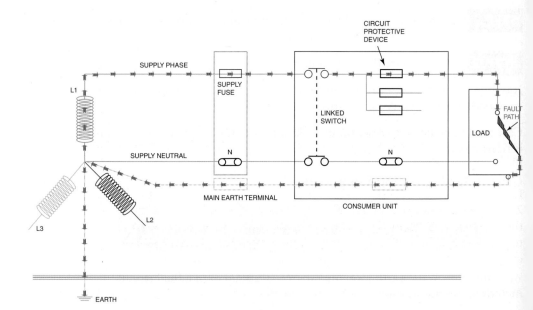

The earth fault loop (Z_S) path for a TNCS system.

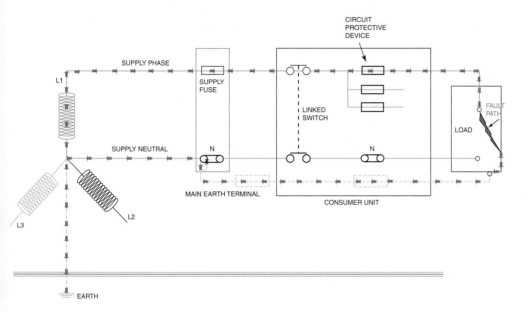

To obtain Z_S, the Z_e value should now be added to the $R_1 + R_2$ values that were obtained when carrying out the CPC continuity tests for each individual circuit.

The total value $Z_e + R_1 + R_2$ will be Z_S (earth loop impedance for the circuit).

This value (Z_S) should now be compared with the maximum values of Z_S given in BS 7671, Chapter 41, to verify that the protective device will operate in the correct time.

Unfortunately, it is not quite as simple as it seems. This is because the values Z_S have been measured when the conductors were at room temperature and the maximum Z_S values given in BS 7671 are at the conductor operating temperature of 70°C. This is the maximum temperature that the conductor could be operating at in a sound circuit. There are two methods.

Accurate test instruments must be used for these tests.

Method one

Measure the ambient temperature of the room and use the values from Table 9B in the *On-Site Guide* as dividers. (**DO NOT USE THEM AS MULTIPLIERS**)

This is because Table 9B is to correct the conductor resistances in Table 9A from 20°C to room temperature.

When they are used as dividers they will correct the cable from room temperature resistance, to the resistance that it would be at 20°C.

Example 7

$$\frac{R_1 + R_2}{\text{Temp factor}}$$

Measured $R_1 + R_2 = 0.84\,\Omega$ @ 25°C

Factor from Table 9b for 25°C = 1.02.

$$\frac{0.84}{1.02} = 0.82$$

Values at 20°C is 0.82 Ω.

Having corrected the measured values to 20°C the next step is to calculate what the resistance of the cable would be at its operating temperature.

As the resistance of copper changes by 2% for each 5°C the conductor resistance will rise by 20% if its temperature rises to 70°C.

Multiplying the resistance by 1.2 will increase its value by 20%, this value can now be added to Z_e to give Z_S.

Resistance value at operating temperature would be:

$$0.82\,\Omega \times 1.2 = 0.98\,\Omega$$

Example 8

The Z_e of an installation is $0.6\,\Omega$. A circuit has been installed using twin and CPC 70°C thermoplastic (pvc) cable. The room temperature is 25°C and the measured $R_1 + R_2$ value is 0.48Ω. The circuit is protected by a BS EN 60898 16A type B device.

Correct the cable resistance to 20°C by using factor from Table 9B *On-Site Guide.*

$$\frac{0.48}{1.02} = 0.47\,\Omega$$

Adjust this value to conductor operating temperature by increasing it by 20%.

$$0.47 \times 1.2 = 0.56\,\Omega$$

Add this value to installation Z_e to find Z_S.

$$0.56 + 0.6 = 1.16\,\Omega$$

This value can now be compared directly with the maximum value Z_S for a 16A type B protective device. This value is $3\,\Omega$ and can be found in Table 41B2 in BS 7671.

To comply with the regulations the actual value $1.16\,\Omega$ is acceptable as it is less than $3\,\Omega$.

Method two

Measure the resistance of earth fault loop impedance at the furthest point of the circuit using the correct instrument (*remember the furthest point is the end of the circuit, not necessarily the furthest distance from the distribution board*). Record the value obtained onto the test result schedule.

This measurement cannot be compared directly with the values from BS 7671 because the operating temperature of the conductors and the ambient temperature of the room are unknown.

This method is useful for a periodic test where existing test results are available. If the measured value is higher than previous results it will indicate that there is a possible deterioration of the earth loop impedance of the circuit.

The usual method to check that the measured Z_S is acceptable is to use the rule of thumb method.

First look in the correct table in Chapter 41 of BS 7671 for the maximum permissible Z_S of the protective device for the circuit being tested.

Use three-quarters of this value and compare it with the measured value. Providing the measured value is the lowest the circuit will comply.

More information on this can be found in Chapter 5 (Protective devices).

Performing the tests

This is a live test and great care must be taken

A circuit incorporating a socket outlet on a ring or a radial

STEP 1 Use an earth fault loop impedance instrument. Set it onto 20Ω (unless you have a self ranging instrument).

STEP 2 Ensure all earthing and bonding is connected.

STEP 3 Plug in the instrument and record the reading.

Performing the test on a radial circuit other than a socket outlet

STEP 1

Ensure earthing and bonding is connected.

STEP 2

Isolate circuit to be tested.

STEP 3

Remove accessory at the extremity of the circuit to be tested.

Probes on phase and earthing
terminals for 2-lead instrument

Two-lead instrument as connected here

STEP 4

Use an earth fault loop impedance instrument with fly leads.

Place the leads on correct terminals. If you are using a two-lead instrument as connected here; if you are using a three-lead instrument as connected here (always read the instrument manufacturers instructions).

STEP 5

Energize the circuit.

Probes on phase and neutral
terminals, clip on earthing terminal
for 3-lead instrument

Three-lead instrumented as connected here

STEP 6

Perform the test on a lighting circuit at the ceiling rose or at the switch.

If you are using a two-lead instrument, place the probes as shown on the left, this will also prove polarity if the switch is operated whilst carrying out the test (*this may be easier with two people*).

Test instrument probe at earthing terminal

Test instrument probe at switched line

Connection to light-fitting place for 2-lead instrument

Test instrument probe at switch return terminal

Test instrument probe at earth terminal

Connection to switch for 2-lead instrument

If you are using a three-lead instrument then connect the probes as shown here (always read the instrument's instructions).

STEP 7 Isolate the circuit and remove the leads.

STEP 8 Replace accessory and energize circuit.

Probe at
switched line

Joined leads at
earthing terminal

Connection to light-fitting place for 3-lead instrument

Probe at switch
return terminal

Joined leads
at earth
terminal

Connection to switch for 3-lead instrument

Example 9

A ring final circuit is protected by a 30A BS3036 semi-enclosed rewirable fuse and the measured Z_S is 0.96 Ω.

As this is a ring final circuit, the disconnection time has to be 0.4 seconds. From Table 41B1 in BS 7671 the maximum Z_S for a 30A rewirable fuse is 1.14 Ω.

Three-quarters of this value must now be calculated, this can be achieved by multiplying it by 0.75.

$$1.14 \times 0.75 = 0.85 \, \Omega$$

The measured value for the circuit must now be lower than the corrected value if it is to comply with BS 7671.

Measured value	*0.96 Ω*
Corrected value	*0.85 Ω*

*The measured value of Z_S is higher; therefore, the circuit will **not** comply.*

Option one is the preferred method because it will give an accurate value whereas the test in Option two will include parallel paths and because of this will often give lower readings.

Option one should always be used for an initial verification as the first reading will be used as a benchmark to be compared with results taken in future periodic tests.

If, on a periodic inspection using Option two, a higher test result is obtained than on the initial verification, this would indicate that the circuit is deteriorating and that further investigation would be required.

The methods described here must be fully understood by anyone who is intending to sit the City and Guilds 2391 exam on inspection and testing of electrical installations.

However! Providing that the ambient temperature is between 10°C and 20°C there is a much simpler way to check the results when you are actually working on site testing. The *On-Site Guide* and *Guidance Note 3* information contain tables which are already corrected for conductor operating current and ambient temperature. In the *On-Site Guide* these tables can be found in Appendix 2 (Table 2A to Table 2D). These tables can also be used where the CPC is of a different cross-sectional area than the live conductors.

These values are slightly more favourable than the rule-of-thumb method, as they are at approximately 80% of values given in BS 7671 and are perfectly acceptable (*remember, however, that the previous methods described <u>must</u> also be understood*).

Example 10

A circuit supplying a fixed load is protected by a 20A BS 3036 fuse and the measured Z_S is 2.32 Ω. The circuit CPC is 1.5 mm^2.

As this circuit is supplying a fixed load the maximum permitted disconnection time is 5 seconds.

The table that should be used is 2A(ii) from the On-Site Guide. Using this table it can be seen that for a 20A device protecting a circuit with a 1.5 mm^2 CPC the maximum permissible Z_S is 3.2 Ω. As the measured Z_S is lower than the maximum permitted, the circuit complies with the Regulations.

Example 11

A circuit supplying a cooker outlet which incorporates a socket outlet protected by a 45A BS 1361 fuse, has a measured Z_S of 0.4 Ω. The CPC is 4 mm^2.

As this circuit has a socket outlet on it, the disconnection time would be 0.4 seconds.

The table to use for this circuit is 2C(i) from the On-Site Guide and the maximum permitted Z_S is 0.48 Ω. This circuit will comply.

Example 12

A circuit supplying a lighting circuit protected by a 6A type C BS 60898 protective device has a measured Z_S of 2.9 Ω.

It should be remembered that miniature circuit breakers will operate at 0.1 seconds providing that the measured Z_S is equal to or lower than the values given in the tables. We do not have to worry about 0.4 or 5 second disconnection times for these devices.

The table to use for this example is 2D in the On-Site Guide, the maximum permitted Z_S is 3.2 Ω. The measured value of 2.9 Ω is lower than the maximum permitted, therefore, this circuit would comply.

Remember that the values in GN3 and the *On-Site Guide* are corrected for temperatures between 10°C and 20°C and no other calculation is required providing the ambient temperature is between these values.

If the ambient temperature is below 10°C or above 20°C then correction factors from Table 2E of the *On-Site Guide* must be used as follows.

Using Example 9, a circuit supplying a fixed load is protected by a 20A BS 3036 fuse and the measured Z_S is 2.32 Ω. The circuit CPC is 1.5 mm², and the **ambient temperature is 23°C.**

As this circuit is supplying a fixed load, the maximum permitted disconnection time is 5 seconds.

The table that should be used is 2A(ii) from the *On-Site Guide*. Using the table it can be seen that for a 20A device protecting a circuit with a 1.5 mm² CPC, the maximum permitted Z_S is 3.20 Ω.

Now the temperature has to be taken into account.

Using Table 2E from Appendix 2 of the *On-Site Guide* it can be seen that the nearest value to the temperature which was measured (23°) is 25°C (*always round up to be on the safe side*). The correction factor for 25°C is 1.06.

This value (1.06) is now used as a multiplier to the maximum permitted Z_S (3.20 Ω) to calculate the maximum Z_S for the circuit at 25°C

$$3.20 \times 1.06 = 3.39 \, \Omega$$

This is the maximum measured Z_S permissible for the circuit at 25°C.

Prospective fault current test

This is a live test and great care must be taken

A prospective fault current test instrument is normally combined with an earth loop impedance test instrument, the measured value is normally shown in kA (kilo amps).

Regulations 434-02-01 and 713-12-01 require that the prospective short circuit current and the prospective earth fault current are determined. Once determined, the highest value must of course be recorded.

A prospective short circuit current is the maximum current that could flow between phase and neutral on a single-phase supply or between phase conductors on a three-phase supply.

A prospective earth fault current is the maximum current that could flow between live conductors and earth.

The higher of these values is known as prospective fault current.

The highest prospective fault current will be at the origin of the installation and must be measured as close to the meter position as possible, usually at the main switch for the installation. It is measured between phase and neutral.

This can be done by:

- Enquiry to the supplier
- Calculation
- Measurement

Enquiry
This is a matter of a phone call to the electricity supplier of the installation. They will tell you the maximum PFC. Usually this is a lot higher than the value will actually be, but if you use this value you will be on the safe side.

Calculation
The PFC can only be calculated on a TNCS system. This is because the neutral of the supply is used as a Protective earth and neutral (PEN) conductor.

When the earth fault loop impedance is measured, the value measured is in ohms. To convert this value to prospective short circuit current we must use the following equation:

$$\text{PSCC} = \frac{V}{Z_e} = I$$

It is important to remember that the open circuit voltage of the supply transformer is used U_{OC} 240 V (BS 7671, Appendix 3).

Example 13

Z_e is measured at $0.28\,\Omega$

$$\frac{240}{0.28} = 857\ amps$$

A useful tip is that when you have measured Z_e on a TNCS system, set your instrument to PFC and repeat the test. This will give you the value for PFC and save you doing the calculation.

Measurement

This is carried out using a prospective fault current tester. As with all tests it is important that you have read the instructions for the instrument which you are going to use.

If you are using a two-lead instrument with leads and probes to GS 38:

STEP 1

Set instrument to PFC.

STEP 2

Place the probes on the phase and neutral terminals at supply side of the main switch.

STEP 3

Operate the test button and record the reading.

When carrying out the test using a three-lead instrument with leads to GS 38, it is important that the instrument instructions are read and fully understood before carrying out this test.

Probe at neutral at supply side of main switch   Insulated tips   Probe at phase at supply side of main switch

Connection for 2-lead instrument

STEP 1

Place the phase lead on the supply side of the main switch and the neutral and earthing probes/clips onto the earthing terminal.

Joined leads at neutral at supply side of main switch   Probe at phase at supply side of main switch

Connection for 3-lead instrument

STEP 2   Operate the test button and record the reading.

If the supply system is a three-phase and neutral system then the highest current that could flow in it will be between phases. Some instruments will not be able to measure the high current that would flow under these circumstances.

Under these circumstances the measurement should be made between any phase and neutral at the main switch and the measured value should be **doubled**.

For your personal safety and the protection of your test equipment it is important to read and fully understand the instructions of your test instrument before commencing this test.

Some PSCC (protective short-circuit current) instruments give the measured value in ohms, not kA. If this is the case, a simple calculation, using ohms law is all that is required.

Example 14

Measured value is $0.08\,\Omega$.

Remember to use U_{OC} in this calculation (240 V)

$$PSCC = \frac{240}{0.08} = 3000\,A$$

It is important that the short circuit capacity of any protective devices fitted exceeds the maximum current that could flow at the point at which they are fitted.

When a measurement of PFC is taken as close to the supply intake as possible, and all protective devices fitted in the installation have a short circuit capacity that is higher than the measured value, then Regulation 432-02-01 will be satisfied.

In a large installation where sub mains are used to supply distribution boards it can be cost effective to measure the PFC at each board. The PFC will be smaller and could allow the use of a protective device with a lower short circuit rating. These will usually be less expensive.

Table 7.2A in the *On-Site Guide* gives rated short circuit capacities for devices. These values can also be obtained from manufacturer's literature.

Examples	Rated short circuit capacity
Semi-enclosed BS 3036	1 kA to 4 kA depending on type
BS 1361 Type 1	16.5 kA
Type 2	33 kA
BS 88-2.1	50 kA at 415 volts
BS 88-6	16.5 kA at 240 volts
	80 kA at 415 volts

Circuit breakers to BS 3871 are marked with values M1 to M9 the number indicates the maximum value of kA that they are rated at.

Circuit breakers to BS EN 60898 and RCBOs to BS EN 61009 show two values in boxes, usually on the front of the device.

Type   In rating   Ics rating   Icn rating

Circuit breaches to BS EN 60898

The square box will indicate the maximum current that the device could interrupt and still be reset $\boxed{3}$. This is the Ics rating. The rectangular box will indicate the maximum current that the device can interrupt safely $\boxed{6000}$. This is the Icn rating.

If a value of fault current above the rated Isc rating of the device were to flow in the circuit, the device will no longer be serviceable and will have to be replaced. A value of fault current above the Icn rating would be very dangerous and possibly result in an explosion causing major damage to the distribution board/consumer's unit.

Functional testing

All equipment must be tested to ensure that it operates correctly. All switches, isolators and circuit breakers must be manually operated to ensure that they function correctly, also that they have been correctly installed and adjusted where adjustment is required.

Residual current device (RCD)

The instrument used for this test is an RCD tester, and it measures the time it takes for the RCD to interrupt the supply of current flowing through it. The value of measurement is either in seconds or milliseconds.

Before we get on to testing, let's consider what types of RCDs there are, what they are used for, and where they should be used.

Types of RCD

Voltage operated

Voltage operated earth leakage current breakers (ELCBs) are not uncommon in older installations. This type of device became obsolete in the early 1980s and must not be installed in a new installation or alteration as they are no longer recognized by BS 7671.

They are easily recognized as they have two earth connections, one for the earth electrode and the other for the installation earthing conductor. The major problem with voltage operated devices is that a parallel path in the system will probably stop it from operating.

These types of devices would normally have been used as earth fault protection in a TT system.

Although the Electrical Wiring Regulations BS 7671 cannot insist that all of these devices are changed, if you have to carry out work on a system which has one it <u>must</u> be replaced to enable certification to be carried out correctly. If, however, a voltage operated device is found while preparing a periodic inspection report, a recommendation that it should be replaced would be the correct way of dealing with it.

BS 4293 General purpose device

These RCDs are very common in installations although they ceased to be used in the early 1990s. They have been replaced by BS EN 61008-1, BS EN 61008-2-1 and BS EN 61008-2-2.

They are used as standalone devices or main switches fitted in consumers' units/distribution boards.

This type of device provides protection against earth fault current. They will commonly be found in TT systems 15 or more years old, although they may be found in TNS systems where greater protection was required.

A typical general purpose RCD to BS4293 – now obsolete

A typical voltage – operated earth leakage current breaker (ELCB) – now obsolete

If this type of device is fitted to a TT system which is being extended or altered, it is quite safe to leave it in the system. If the system supplies socket outlets which could be used to supply portable equipment used outside, it must have a tripping current of no more than 30 mA. This includes socket outlets that could serve extension leads passed through open windows or doors.

The problem with using a low tripping current device as the main switch is that nuisance tripping could occur. The modern way of tackling this is explained later in this chapter.

BS 4293 Type S

These are time delayed RCDs and are used to give good discrimination with other RCDs.

BS EN 61008-1 General purpose device

This is the current standard for a residual current circuit breaker (RCCB) and provides protection against earth fault current. These devices are generally used as main switches in consumers' units/distribution boards.

Three-phase devices are also very common.

BS 7288

This is the current standard for RCD-protected socket outlets and provides protection against earth fault currents. These socket outlets would be used in areas where there is an increased risk of electric shock, such as common areas of schools and colleges. It is also a requirement that any socket outlet used for portable equipment outdoors must have supplementary protection provided by an RCD (*Regulation 471-16-02*). Where the socket outlets are sited outside, waterproof BS 7288 outlets are used to IP 56.

A typical 2-phase
RCCB to BS EN 61008-1

A typical 3-phase RCCB to BS EN 61008-1

BS EN 61009-1

This is the standard for a residual current circuit breaker with overload (RCCBO) protection.

These devices are generally used to provide single circuits with earth fault protection, overload protection and short circuit protection. They are fitted in place of miniature circuit breakers and the correct type should be used (types B, C or D).

BS EN 61008-1 Type S

These are time delayed RCDs and are used to give good discrimination with other RCDs.

Section 3 of the *On-Site Guide* gives good examples of how these devices should be used within an installation.

RCDs and supply systems

TT system

If the installation is a TT system and it is possible for any of the sockets to supply portable equipment outdoors or RCD protection is required for other

A typical RCCBO to BS EN 61009-1

Side view of a typical RCCBC to BS EN 61009-1

reasons (*protection for fixed equipment in zones 1-3 in bathrooms, for instance*), there are various options available. While all will be safe, they will vary in cost.

Option 1 Use a 100 mA S-Type RCD (*BS EN 610081*) as the main switch and RCD protection for all circuits. Then use a 30 mA RCBO (*BS EN 61009*) as a circuit protective device for the circuit which requires protection. In this case, the 100 mA RCD must be labelled 'Main Switch'.

Option 2 Use a split board with 100 mA S-Type RCD (*BS EN 61008*) as a main switch and a 30 mA RCD (*BS EN61008*). All circuits require supplementary protection. This method is useful if more than one circuit requires 30 mA protection.

100 mA S-type   30 mA RCD   Only this part is
RCD (BS EN 61008)   (BS EN 61008)   RCD – protected
as main switch   protection

Option 3 Use a split board with a 100 mA RCD for fixed equipment and lighting with a 30 mA RCD for the circuits requiring supplementary bonding. This option would require a separate main switch.

Option 4 Another method would be to use a consumer unit with a main switch to BSEN 60947-3 and RCBOs to BS EN 610091 as protective devices for all circuits. This option is perfectly satisfactory but can work out a little expensive!

TNS and TNCS systems

If the supply system was TNS or TNCS, Options 1 and 2 could be used. But instead of using a 100 mA S-Type RCD as a main switch, it could be substituted for a main switch to BS EN 60947-3.

Option 4 will remain pretty much the same but BS EN 60898 devices could be used where RCD protection is not required.

Testing of RCDs

> **Remember that these are live tests and care should be taken whilst carrying them out**

The instrument to be used to carry out this test is an RCD tester, with leads to comply with GS 38.

Voltage operated (ELCBs)

No test required as they should now be replaced.

BS 4293 RCDs

If this type of RCD is found on TT systems or other systems where there is a high value of earth fault impedance (Z_e), the RCD tester should be plugged into the nearest socket or connected as close as possible to the RCD. The tester should then be set at the rated tripping current of the RCD ($I\Delta n$); for example, at 30 mA (*be careful and do not mistake the tripping current for the current rating of the device*).

Before carrying out the tests ensure that all loads are removed; failure to do this may result in the readings being inaccurate.

STEP 1

The test instrument must then be set at 50% of the tripping current (15 mA).

STEP 2

Push the test button of the instrument, the RCD should not trip.

Switch on 180°   Set to rated tripping current of the RCD   Set at 50% of the tripping current

STEP 3

The test instrument will have a switch on it which will enable the instrument to test the other side of the waveform 0°~180°. This switch must be moved to the opposite side and the test repeated. Again the RCD should not trip.

Switch on 0°   Set to rated tripping current of the RCD   Set at 50% of the tripping current

If, while testing an RCD it trips during the 50% test, do not automatically assume that the RCD is at fault.

Consider the possibility that there is a small earth leakage on the circuit or system. Switch all circuits off and test RCD on the load side at 50% using fly leads. If it still trips, then the RCD should be replaced.

If the RDC does not trip, then turn each circuit on one at a time, carrying out a 50% test each time a circuit has been turned on. When the RCD does trip, switch off all circuits except the last one which was switched on. Test again. If the RCD trips carry out an insulation test on this circuit as it probably has a low insulation resistance. If, however, the RCD does not trip it could be an accumulation of earth leakage from several circuits and they should all be tested for insulation resistance.

STEP 4

Now set the test current to the rated tripping current (30 mA)

STEP 5

Push the test button, and the RCD should trip within 200 milliseconds.

STEP 6

Reset the RCD.

STEP 7

Move the waveform switch to the opposite side, and repeat the test. Again it must trip within 200 milliseconds.

STEP 8

Reset the RCD and the slowest time in which it trips should be entered on to the test result schedule.

If the RCD is used to provide supplementary protection against direct contact, or for socket outlets used to supply portable equipment outdoors – fixed equipment in zone 1 or other current using equipment in zone 3 of a bathroom, the following test should be performed.

| Switch on 0° | Tripping time <200 ms | Set to rated tripping current of the RCD | Set to 100% of the tripping current |

| Switch on 180° | Tripping time <200 ms | Set to rated tripping current of the RCD | Set to 100% of the tripping current |

STEP 9

Set the test current to 5 times the rated tripping current (150 mA).

STEP 10

Push the test button and the RCD should trip within 40 milliseconds.

| Switch on 180° | Tripping time <40ms | Set to rated tripping current of the RCD | Set to 5 times the rated tripping current |

STEP 11

Move the waveform switch to the opposite side, and repeat the test. Again it must trip within 40 milliseconds (five times faster than the times 1 test).

After completion of the instrument tests

STEP 12

Push integral test button on RCD to verify that the mechanical parts are working correctly.

STEP 13

Ensure that a label is in place to inform the user of the necessity to use the test button quarterly.

| Switch on 0° | Tripping time <40 m | Set to rated tripping current of the RCD | Set to 5 times the rated tripping current |

BS EN 610081

These devices should be tested in exactly the same manner as BS 4293 using the same test instrument. However, the difference is that, when carrying out the 100% test, the tripping time is increased to 300 milliseconds.

BS 4293 Type S

This device has a built in time delay. The simple way to think about this is that it does not recognize a fault for 200 milliseconds, and they must trip within 200 milliseconds after that.

STEP 1   Plug in or connect an RCD as close as possible to the RCD to be tested.

STEP 2   Set the instrument on the trip current of the RCD and ensure that it is set for 'S' type.

STEP 3   Test at 50% and the device should not trip.

STEP 4   Repeat the test on the opposite waveform.

STEP 5   Set test instrument on 100% and carry out test. The RCD should trip within 400 milliseconds (200 *ms time delay and 200 ms fault*).

STEP 6   Repeat on the opposite wave form.

The slowest operating time at the 100% test should be recorded, as should the fact that it is an 'S' type.

BS EN 61008 Type S

This device has a time delay of 200 milliseconds and a tripping time of 300 milliseconds, making a maximum tripping time of 500 milliseconds.

The test should be carried out as the BS 4293 Type S but remember the different tripping time.

BS 7288 RCD protected socket

This device should be tested the same as for a BS 4293 and the tripping times are the same.

The 5 times test must only be carried out on RCDs with trip ratings (I∆n) up to 30 mA.

Always ensure that it is safe to carry out these tests. Remember to remove any loads, ensure that the disconnecting of the supply due to the test will not effect any equipment or cause damage. If any people are within the building ensure that they are aware of testing being carried out, and that a loss of supply is likely.

Consideration should be given to whether the socket will supply portable equipment outdoors. If it can it should be tested at 5 times its rating.

BS EN 61009 RCBOs

These devices should be tested as for BS 4239 RCDs but the disconnection times are:

- 50% test on both sides of waveform, no trip.
- 100% test on both sides of waveform; must trip within 300 milliseconds.
- If used as supplementary protection, the 5 times test must also be carried out; it must trip within 40 milliseconds.

4

Completion of test certificates

The following pages detail the test certificates and itemized descriptions. At the end of the chapter further 'Notes for Recipients' can be found.

Minor Domestic Electrical Installation Works Certificate

This certificate is to be completed when additions to existing circuits have been carried out. For example, an additional lighting point or socket outlet.

If more than one circuit has been added to, then a separate Minor Works Certificate must be issued for each modification.

These certificates vary slightly depending on which certification body has supplied them; some require slightly more information than others. The information required is as follows.

DOMESTIC INSTALLER

This safety certificate is an important and valuable document which should be retained for future reference

DMP2/

TRACEABLE SERIAL NUMBER

Original *(To the person ordering the work)*

MINOR DOMESTIC ELECTRICAL INSTALLATION WORKS CERTIFICATE

Issued in accordance with *British Standard 7671 – Requirements for Electrical Installations* by an Approved Contractor or Conforming Body enrolled with NICEIC, Warwick House, Houghton Hall Park, Houghton Regis, Dunstable, LU5 5ZX.

To be used only for minor electrical work which does not include the provision of a new circuit

PART 1: DETAILS OF THE MINOR WORKS

Client

Date minor works completed

Contract reference, if any

Description of the minor works

Details of departures, if any, from BS 7671 (as amended)

Location/address of the minor works

Postcode

PART 2: DETAILS OF THE MODIFIED CIRCUIT

System type and earthing arrangements   TN-C-S   TN-S   TT

Method of protection against indirect contact   EEBAD   Other

Overcurrent protective device for the modified circuit   BS(EN)   Type   Rating   A

Residual current device (if applicable)   BS(EN)   Type $I_{\Delta n}$ mA

Details of wiring system used to modify the circuit   Type   Reference method   csa of live conductors mm^2 csa of cpc mm^2

Where protection against indirect contact is EEBAD,   Maximum disconnection time permitted by BS 7671   s   Maximum Z_s permitted by BS 7671 Ω

Comments, if any, on existing installation

PART 3: INSPECTION AND TESTING OF THE MODIFIED CIRCUIT AND RELATED PARTS   † *Essential inspections and tests*

† Confirmation that necessary inspections have been undertaken	(✓)	† Confirmation of the adequacy of earthing	(✓)
† Circuit resistance $R_1 + R_2$ Ω or R_2 Ω		† Confirmation of the adequacy of equipotential bonding	(✓)
Insulation resistance (* *In a multi-phase circuit, record the lower or lowest value, as appropriate*) Phase/Phase* $M\Omega$		† Confirmation of correct polarity	(✓)
Phase/Neutral* $M\Omega$		† Maximum measured earth fault loop impedance, Z_s	Ω
Instrument Serial No(s). † Phase/Earth* $M\Omega$		† RCD operating time at $I_{\Delta n}$ (if RCD fitted)	ms
† Neutral/Earth $M\Omega$		RCD operating time at 5 x $I_{\Delta n}$, if applicable	ms

Agreed limitations, if any, on the inspection and testing

PART 4: DECLARATION

I/We CERTIFY that the said works do not impair the safety of the existing installation, that the said works have been designed, constructed, inspected and tested in accordance with BS 7671:   (IEE Wiring Regulations), amended to   and that the said works, to the best of my/our knowledge and belief, at the time of my/our inspection complied with BS 7671 except as detailed in Part 1.

For and on behalf of *(Trading Title of Domestic Installer)*

The results of the inspection and testing reviewed by the Qualified Supervisor

Name (CAPITALS)	Name (CAPITALS)	Address
Signature	Signature	
Date	Date	Postcode
Registration Number		*(The registration number is essential information)*

This form is based on the model shown in Appendix 6 of BS 7671 (as amended)
Published by the NICEIC Group Limited © Copyright The Electrical Safety Council (July 2006)

Please see the 'Notes for Recipients' on the reverse of this page.

DMP2/1

Client Name of the person ordering the work.

Location/address The address at which the work is carried out.

Date Completion of minor work.

Description It is important to document exactly the work which has been carried out.

Type of system TT, TNS, TNCS.

Method of protection against indirect contact This will usually be EEBAADS.

Overcurrent device for the circuit This is the type and size of device which is protecting the circuit on which the minor works has been carried out. If it is necessary to change the protective device, then an electrical installation certificate is required. Not a Minor Works Certificate.

Residual current device Type, current rating and tripping value IΔn is required. If the addition to the circuit requires the fitting of an RCD, then an Electrical Installation Certificate is required for the RCD.

Details of wiring system used What type of wiring is it? For example, PVC, conduit, steel wire armour.

Reference method How has it been installed? See Appendix 4, BS 7671 for reference methods.

CSA of conductors What size are the conductors?

Maximum disconnection time of the circuit Is it 5 seconds or 0.4 seconds?

Maximum Z_S What is the maximum Z_S permitted to ensure that the protective device operates in the correct time? These can be found in Part 4 of BS 7671.

Circuit resistance What is the value of $R_1 + R_2$ or R_2 if Table 41C of BS 7671 is being used.

Confirmation of bonding Has bonding been installed? If not, this should be pointed out to the client, and identified in the comments section of the certificate. *It is still permissible to carry out minor works if bonding is not present in the installation.*

Confirmation of earthing Is the installation earthed? If not, then the work should not be carried out and the client should be informed of the danger that this presents.

Correct polarity Is the supply correct? Have the circuit conductors been connected in to the correct terminals?

Measured Z_S What is the measured value of Z_S for the altered circuit? Check that it is lower than the maximum permitted value.

RCD operating time at $I\Delta n$ Must be less than 300 ms if BS EN type, or 200 ms if BS type.

RCD operating time at $5\ I\Delta n$ Only required for RCDs rated at 30 mA or less, when used for supplementary protection against direct contact. Never required on RCDs above 30 mA.

Insulation resistance Values required for between live conductors and live conductors and earth. If the measured value is below $2\ M\Omega$, further investigation is required. For 3 phase circuits record the lowest value.

Comments on existing installation Generally, just a comment on the visual condition of the installation; such as, is it old? Perhaps a periodic inspection report may be advised. Is the earthing up to the current requirements necessary for BS 7671?

Agreed limitations on the inspecting and testing Not usually many on a minor works. Could possibly be where it is difficult to disconnect or isolate vulnerable equipment and the insulation resistance test is carried out between live conductors joined and earth only.

NOTES FOR RECIPIENTS

THIS SAFETY CERTIFICATE IS AN IMPORTANT AND VALUABLE DOCUMENT
WHICH SHOULD BE RETAINED FOR FUTURE REFERENCE

IF YOU WERE THE PERSON ORDERING THE WORK, BUT NOT THE OWNER OR USER OF THE INSTALLATION, YOU SHOULD PASS THIS CERTIFICATE, OR A FULL COPY OF IT INCLUDING THESE NOTES, IMMEDIATELY TO THE OWNER OR USER OF THE INSTALLATION.

This safety certificate has been issued to confirm that the minor electrical installation works to which it relates has been designed, constructed, inspected, tested and verified in accordance with the national standard for the safety of electrical installations, British Standard 7671 (as amended) - *Requirements for Electrical Installations* **(the IEE Wiring Regulations).**

Where, as will often be the case, the existing installation incorporates a residual current device (RCD), there should be a notice at or near the main switchboard or consumer unit stating that the device should be tested at quarterly intervals. For safety reasons, it is important that you carry out the test regularly.

Also, for safety reasons, the complete electrical installation including the minor electrical installation works which is the subject of this certificate will need to be inspected and tested at appropriate intervals by a competent person. NICEIC* recommends that you engage the services of an NICEIC Approved Contractor for this purpose. There should be a notice at or near the origin of the existing installation (such as at the consumer unit or main switchboard) which indicates when the inspection of the complete installation is next due.

Only the NICEIC Domestic Installer responsible for the work is authorised to issue this NICEIC certificate. The certificate has a printed serial number which is traceable to the Domestic Installer to which it was supplied by NICEIC.

You should have received the certificate marked 'Original' and the Domestic Installer should have retained the certificate marked 'Duplicate'. **The 'Original' certificate should be retained in a safe place and shown to any person inspecting, or undertaking further work on, the electrical installation in the future. If you later vacate the property, this certificate will demonstrate to the new user that the minor electrical installation works complied with the requirements of the national electrical safety standard at the time the certificate was issued.**

The Minor Domestic Electrical Installation Works Certificate is intended to be used only for an addition or alteration to an existing circuit that does not extend to the provision of a new circuit. Examples include the addition of a socket-outlet to an existing circuit or the addition of a lighting point to an existing circuit, or the replacement or relocation of a light switch. A separate certificate should have been received for each existing circuit on which minor works has been carried out. This certificate would be considered by NICEIC to be invalid if you requested the Domestic Installer to undertake more extensive work for which a Domestic Electrical Installation Certificate should have been issued.

Part 3 of this certificate is intended to facilitate the recording of information associated with the inspection and testing of the modified circuit, and the related parts of the existing installation on which the modified circuit depends for its safety. Generally, each box should have been completed to confirm the results of a particular inspection or test by a 'Yes' or a '✓', or by the insertion of a measured value. Where a particular inspection or test was not applicable, this should have been indicated by 'N/A', meaning 'Not Applicable'. Where an inspection or a test was not practicable, the entry should read 'LIM', meaning 'Limitation', acknowledging that the particular circumstances prevented the particular inspection or test procedure from being carried out. In such a case, each limitation should have been recorded in the box entitled 'Agreed limitations, if any, on the inspection and testing', together with the reason for each limitation.

If wiring alterations or additions are made to an installation such that wiring colours to two versions of BS 7671 exist, a warning notice should have been affixed at or near the appropriate consumer unit.

Should the person ordering the work (eg the client, as identified on this certificate), have reason to believe that any element of the work for which the Domestic Installer has accepted responsibility (as indicated by the signature on this certificate) does not comply with the requirements of the national electrical safety standard (BS 7671), the client should in the first instance raise the specific concerns in writing with the Domestic Installer. If the concerns remain unresolved, the client may make a formal complaint to NICEIC, for which purpose a standard complaint form is available on request.

The complaints procedure offered by NICEIC is subject to certain terms and conditions, full details of which are available upon application and from the website[†]. NICEIC does not investigate complaints relating to the operational performance of electrical installations (such as lighting levels), or to contractual or commercial issues (such as time or cost).

* NICEIC' is a trading name of NICEIC Group Limited, a wholly owned subsidiary of The Electrical Safety Council. Under licence from The Electrical Safety Council, NICEIC acts as the electrical contracting industry's independent voluntary regulatory body for electrical installation safety matters throughout the UK, and maintains and publishes registers of electrical contractors that it has assessed against particular scheme requirements (including the technical standard of electrical work).

For further information about electrical safety and how the NICEIC can help you, visit **www.niceic.com**

Electrical Installation Certificate

This certificate is to be completed for a new circuit, a new installation, a rewiring and any circuit where the protective device has been changed.

In the case of a consumer's unit change only, an Electrical Installation Certificate would be required for the consumers unit and a Periodic Inspection Report should be completed for the existing installation.

This certification is required for a new installation or circuit

A standard Electrical Installation Certificate can be used for any installation. However, if the work to be certificated is covered by building regulation, a Part P Certificate is available solely for this purpose. These certificates simplify the paper work by including a schedule of inspection and a Schedule of Test Results on the same document.

A Schedule of Test Results and a Schedule of Inspection must be completed to accompany an electrical installation certificate and a periodic inspection report.

These certificates vary slightly depending on which certification body has supplied them, some require slightly more information than others. The Electrical Installation Certificate and Particulars of Signatures to the Electrical Installation Certificate are typical of an electrical installation certificate. The information required is as follows.

This safety certificate is an important and valuable document which should be retained for future reference

ELECTRICAL INSTALLATION CERTIFICATE

Issued in accordance with *British Standard BS 7671- Requirements for Electrical Installations*

DETAILS OF THE CLIENT

Client / Address:

DETAILS OF THE INSTALLATION

The installation is:

Address:

New

Extent of the installation covered by this certificate:

An addition

An alteration

DESIGN

I/We, being the person(s) responsible for the design of the electrical installation (as indicated by my/our signature(s) below), particulars of which are described above, having exercised reasonable skill and care when carrying out the design, hereby CERTIFY that the design work for which I/we have been responsible is to the best of my/our knowledge and belief in accordance with BS 7671: _____ amended to _____ (date) except for the departures, if any, detailed as follows:

Details of departures from BS 7671, as amended (Regulations 120-01-03, 120-02):

The extent of liability of the signatory/signatories is limited to the work described above as the subject of this certificate.
For the **DESIGN** of the installation: **(Where there is divided responsibility for the design)**

| Signature | | Date | | Name (CAPITALS) | | Designer 1 |
| Signature | | Date | | Name (CAPITALS) | ** | Designer 2 |

CONSTRUCTION

I/We, being the person(s) responsible for the construction of the electrical installation (as indicated by my/our signature below), particulars of which are described above, having exercised reasonable skill and care when carrying out the construction, hereby CERTIFY that the construction work for which I/we have been responsible is to the best of my/our knowledge and belief in accordance with BS 7671: _____ amended to _____ (date) except for the the departures, if any, detailed as follows:

Details of departures from BS 7671, as amended:

The extent of liability of the signatory is limited to the work described above as the subject of this certificate.
For the **CONSTRUCTION** of the installation:

| Signature | | Date | | Name (CAPITALS) | | Constructor |

INSPECTION AND TESTING

I/We, being the person(s) responsible for the inspection and testing of the electrical installation (as indicated by my/our signatures below), particulars of which are described above, having exercised reasonable skill and care when carrying out the inspection and testing, hereby CERTIFY that the work for which I/we have been responsible is to the best of my/our knowledge and belief in accordance with BS 7671: _____ amended to _____ (date) except for the departures, if any, detailed as follows:

Details of departures from BS 7671, as amended:

The extent of liability of the signatory/signatories is limited to the work described above as the subject of this certificate.
For the **INSPECTION AND TESTING** of the installation: Reviewed by †

| Signature | | Date | | Signature | | Date |
| Name (CAPITALS) | | Inspector | Name (CAPITALS) | | |

DESIGN, CONSTRUCTION, INSPECTION AND TESTING *

* This box to be completed only where the design, construction, inspection and testing have been the responsibility of one person.

I, being the person responsible for the design, construction, inspection and testing of the electrical installation (as indicated by my signature below), particulars of which are described above, having exercised reasonable skill and care when carrying out the design, construction, inspection and testing, hereby CERTIFY that the said work for which I have been responsible is to the best of my knowledge and belief in accordance with BS 7671, _____ amended to _____ (date) except for the departures, if any, detailed as follows:

Details of departures from BS 7671, as amended (Regulations 120-01-03, 120-02):

The extent of liability of the signatory is limited to the work described above as the subject of this certificate.
For the **DESIGN**, the **CONSTRUCTION** and the **INSPECTION AND TESTING** of the installation.   Reviewed by †

| Signature | | Date | | Signature | | Date |
| Name (CAPITALS) | | | Name (CAPITALS) | | |

† The completed schedules of inspection and testing should preferably be reviewed by another competent person to confirm that the recorded results are consistent with electrical installation work conforming to the requirements of BS 7671: 2001.

Page 1 of _____

This form is based on the model shown in Appendix 6 of BS 7671: 2001.
© Copyright Electrical Safety Council (July 2006).

Please see the 'Notes for Recipients' on the reverse of this page.

ICM3/1

Details of client Name and address of the person ordering the work.

Location/address The address at which the work is carried out.

Details of the installation What part of the installation does this certificate cover: is it all of the installation, or is it a single circuit? It is vital that this part of the certificate is completed as accurately as possible.

There are generally three tick boxes regarding the nature of the installation.

New To be ticked if the whole installation is new. This would include a rewire.

Alteration To be ticked where the characteristics of an existing circuit have been altered (such as extending/altering a circuit and changing the protective device). This box would also cover the replacement of consumer units and/or the fitting of RCDs.

Addition Used to identify when a new circuit or numerous circuits have been added to an existing circuit.

Design, construction, inspection and testing The person or persons responsible for each of these must sign. It could be one person or possibly three, depending on the job. However it is important that all boxes have a signature.

Usually in this section there will be two boxes referring to BS 7671. To complete this correctly, look at the top right-hand corner of BS 7671 – where the words 'BS 7671 "Year"' will be seen. Just below this will be the date of the amendments. This will indicate the most recent amendment.

Next inspection The person who has designed the installation, or the part of it that this certificate covers must recommend when the first periodic inspection and test is carried out. This will be based on the type of use to which it will be put, and the type of environment.

PARTICULARS OF THE ORGANISATION(S) RESPONSIBLE FOR THE ELECTRICAL INSTALLATION

DESIGN (1)   Organisation †

Address:

Postcode

DESIGN (2)   Organisation †

Address:

Postcode

CONSTRUCTION †   Organisation

Address:

Postcode

INSPECTION AND TESTING   Organisation †

Address:

Postcode

SUPPLY CHARACTERISTICS AND EARTHING ARRANGEMENTS   *Tick boxes and enter details, as appropriate*

✧ System Type(s)	✧ Number and Type of Live Conductors		Nature of Supply Parameters			✧ Characteristics of Primary Supply Overcurrent Protective Device(s)	
TN-S	a.c.	d.c.	Nominal voltage(s): $U^{(1)}$	V $U_o^{(1)}$ V	Notes:		
TN-C-S	1-phase (2 wire)	1-phase (3 wire)	2 pole	Nominal frequency, $f^{(1)}$ Hz	(1) by enquiry	BS(EN)	
TN-C	2-phase (3 wire)		3-pole	Prospective fault current, $I_{pf}^{(2)(3)}$ kA	(2) by enquiry or by measurement	Type	
TT	3-phase (3 wire)	3-phase (4 wire)	other	External earth fault loop impedance, $Z_e^{(2)(3)}$ Ω	(3) where more than one supply, record the higher or highest values	Nominal current rating	A
IT	Other *Please state*			Number of supplies		Short-circuit capacity	kA

PARTICULARS OF INSTALLATION AT THE ORIGIN   *Tick boxes and enter details, as appropriate*

✧ Means of Earthing		Details of Installation Earth Electrode (where applicable)	
Supplier's facility:	Type: (eg rod(s), tape etc)	Location:	
Installation earth electrode:	Electrode resistance, R_A: (Ω)	Method of measurement:	

✧ Main Switch or Circuit-Breaker
* *(applicable only where an RCD is suitable and is used as a main circuit-breaker)*

		Maximum Demand (Load):	A per phase	Method of Protection against Indirect Contact:		

		Main Protective Conductors				
Type: BS(EN)	Voltage rating V	Earthing conductor	Main equipotential bonding conductors	Bonding of extraneous-conductive-parts (✓)		
No of Poles	Current rating, I_n A	Conductor material	Conductor material	Water service	Gas service	
Supply conductors: material	RCD operating current, $I_{\Delta n}$* mA	Conductor csa mm²	Conductor csa mm²	Oil service	Structural steel	
Supply conductors: csa mm²	RCD operating time (at $I_{\Delta n}$)* ms	Continuity check (✓)	Continuity check (✓)	Lightning protection	Other incoming service(s)	

COMMENTS ON EXISTING INSTALLATION

Note: Enter 'NONE' or, where appropriate, the page number(s) of additional page(s) of comments on the existing installation.

NEXT INSPECTION   § *Enter interval in terms of years, months or weeks, as appropriate*   §

I/We, the designer(s), RECOMMEND that this installation is further inspected and tested after an interval of not more than

†   *Where the electrical contractor responsible for the construction of the electrical installation has also been responsible for the design **and** the inspection and testing of that installation, the 'Particulars of the Organisation Responsible for the Electrical Installation' may be recorded only in the section entitled 'CONSTRUCTION'.*

✧   *Where a number of sources are available to supply the installation, and where the data given for the primary source may differ from other sources, a separate sheet must be provided which identifies the relevant information relating to each additional source.*

Page 2 of ☐

This form is based on the model shown in Appendix 6 of BS 7671: 2001.
© Copyright Electrical Safety Council (July 2006).

ICM3/3

Supply characteristics

Type of system Is it TT, TNS or TNCS?

Number and type of live conductors Usually 1 phase 2 wire or 3 phase 4 wires (only live conductors).

Nature of supply parameters This can be gained by enquiry or measurement.

U. Is phase to phase or phase
U_o Is phase to earth

Do not record measured values. If 3 phase or 3 phase and neutral, then the value will be U. 400 V and U_o 230 V. If single phase, then U and U_o will be 230 V.

External earth loop impedance (Z_e) It should be measured between the phase and earth on the live side of the main switch with the earthing conductor disconnected (remember *to isolate the installation first*). If measurement is not possible, then it can be obtained by asking the supply provider.

Overcurrent protective device for the supply Usually this will be a BS 88 or BS 1361 cartridge fuse and it will normally be marked on the supply cut out. If it is not, then it should be found by asking the supply provider.

It is important that the BS or BS EN number, type, current rating and short circuit capacity is recorded (*short circuit capacities can be found in Table 7.2A in the On-Site Guide*).

Particulars of installation at the origin

Means of earthing Is the earthing supplied by the distributor or has it got an earth electrode?

Details of earth electrode If the system has an earth electrode, what type is it? Where is it? What is its resistance? (*This is usually measured with an earth loop impedance tester, using the same method as for Z_e.*)

Maximum demand What is the load per phase? This value is not the rating of the supply fuse or the addition of the circuit protective device ratings. It must be assessed using diversity. The use of Table 1B in Appendix 1 of the *On-Site Guide* can be helpful. However, the use of common sense and experience is probably the best way to deal with this.

Earthing conductor Conductor material, what material is the earthing conductor made of? It must be copper if less than $16\,mm^2$.

Conductor CSA Size of the conductor (in mm^2).

Continuity This must be measured end to end or, if it is visible for its entire length, a visual check is OK.

Main equipotential bonding conductors What are they made of and what has been bonded?

Main switch or circuit breaker What type of switch is it – BS, BS EN or IEC?

Voltage and current rating Whatever is marked on it.

Number of poles Single pole, double pole, triple pole or triple pole and neutral.

Supply conductors Size of tails (in mm^2).

If an RCD is fitted as a <u>main switch</u>, the operating current $I\Delta n$ and the operating time at $I\Delta n$ must be recorded.

Comments on the existing installation If the certificate covers the whole installation then usually 'none' will be entered here. If the installation is old, or you have any concerns, you may enter here that a Periodic Inspection Report would be advisable. Perhaps there are socket outlets which could be used for the supply of portable equipment used outside. In this instance, a recommendation that an RCD is fitted may be entered.

Wiring regulations are not retrospective, so it is not a requirement when wiring that complied when it was first installed is then updated. As you may be the first person to have looked at the installation for many years, your professional advice should be important to your client, and it may be an excellent sales opportunity. It is important to remember that if you are completing an Electrical Installation Certificate, then the earthing and bonding arrangements must be improved to comply with the requirements set out by BS 7671.

Schedule of Circuit Details and Test Results

The Schedule of Test Results and the Schedule of Circuit Details are the basic documents. Some certification bodies have certificates which are a little more comprehensive. Details of how to complete them can be found in the Survey and Test Results Schedule – it is important that when completing this document that each box in the row being completed is marked (this can be the value, a tick, 'n/a' or '/').

SCHEDULE OF CIRCUIT DETAILS
FOR THE INSTALLATION

Original (To the person ordering the work)

TO BE COMPLETED IN EVERY CASE	ONLY TO BE COMPLETED IF THE DISTRIBUTION BOARD IS NOT CONNECTED DIRECTLY TO THE ORIGIN OF THE INSTALLATION*		
Location of distribution board:	Supply to distribution board is from:	No of phases:	Nominal voltage: V
Distribution board designation:	Overcurrent protective device for the distribution circuit: Type: BS(EN):	Associated RCD (if any): BS(EN)	
	Rating: A	RCD No of poles:	I$_{\Delta n}$ mA

CIRCUIT DETAILS

Circuit number and phase	Circuit designation	Type of wiring (see code below)	Reference method †	Number of points served	Circuit conductors: csa Live (mm²)	Circuit conductors: csa cpc (mm²)	Max. disconnection time permitted by BS 7671 (s)	Overcurrent protective devices BS (EN)	Type No	Rating (A)	Short-circuit capacity (kA)	RCD Operating current I$_{\Delta n}$ (mA)	Maximum Z$_s$ permitted by BS 7671 (Ω)

† See Table 4A1 of Appendix 4 of BS 7671.

CODES FOR TYPE OF WIRING									
A	B	C	D	E	F	G	H	0 (Other - please state)	
PVC/PVC cables	PVC cables in metallic conduit	PVC cables in non-metallic conduit	PVC cables in metallic trunking	PVC cables in non-metallic trunking	PVC/SWA cables	XLPE/SWA cables	Mineral-insulated cables		Page 4 of

* In such cases, details of the distribution (sub-main) circuit(s), together with the test results for the circuit(s), must also be provided on continuation schedules.

This form is based on the model shown in Appendix 6 of BS 7671: 2001.
© Copyright Electrical Safety Council (July 2006).

See next page for
Schedule of Test Results

ICM3/7

Location of distribution board Where is it?

Designation of consumer unit If there is more than one unit how is it identified – number, name or letter?

Circuit designation What does the circuit supply? Is it a cooker, ring, etc?

If the circuit is fed by a submain (*distribution circuit*), details of the sub-main must be recorded. Possibly on a separate schedule, or on the top line of the schedule which you are completing.

Type of wiring Is it PVC twin and CPC, plastic or steel conduit? Some certification bodies have their own codes for this.

Reference method How is it installed? The methods are detailed in Appendix 4 of BS 7671.

Number of points served How many outlets or items of fixed equipment are on the circuit?

Circuit conductor size What size are live conductors and CPC? (give in mm^2).

Maximum disconnection time permissible for the circuit Normally 0.4 seconds for circuits supplying hand held equipment; 5 seconds for fixed equipment. Some special locations will have other requirements.

Type of over-current protective device

BS or BS EN. Enter the number.

Type If BS 1361 it will be a number 1, 2, 3 or 4; if BS EN it will be B, C, or D. The letter A was not used in case it was mistaken for Amperes. If the device is a fuse '/' should be entered.

Rating What is the current rating of the device?

Short circuit capacity What is the short circuit capacity of the device? It may be marked on the device; if it is not, then Table 7.2A in the *On-Site Guide* will be of assistance.

Whatever it is, it must be at least equal to the PFC measured at the main switch of the consumer unit.

Maximum Z_S

This is the maximum Z_S permitted by BS 7671 for the protective device in this circuit. This value can be found in Tables 41B1, 41B2 or 41D in the *On-Site Guide*. Be careful to use the correct table relating to the disconnection time for the circuit if fuses are used. Do not use Table 41C here.

SCHEDULE OF TEST RESULTS
FOR THE INSTALLATION

ONLY TO BE COMPLETED IF THE DISTRIBUTION BOARD IS NOT CONNECTED DIRECTLY TO THE ORIGIN OF THE INSTALLATION

★ See note below | Characteristics at this distribution board

Z_s ★   Ω   Operating times of associated RCD (if any)   At $I_{\Delta n}$ ms

I_{pf} ★   kA   At $5 I_{\Delta n}$ (if applicable)   ms

Test instruments (serial numbers) used:

Earth fault loop impedance		RCD	
Insulation resistance		Other	
Continuity		Other	

TEST RESULTS

Circuit number and phase	Circuit impedances (Ω)					Insulation resistance † *Record lower or lowest value*				Polarity	Maximum measured earth fault loop impedance, Z_s ★ *See note below*	RCD operating times	
	Ring final circuits only (measured end to end)			All circuits (At least one column to be completed)		Phase/Phase †	Phases/Neutral †	Phases/Earth †	Neutral/Earth			at $I_{\Delta n}$	at $5 I_{\Delta n}$ (if applicable)
	r_1 (Phase)	r_n (Neutral)	r_2 (cpc)	$R_1 + R_2$	R_2	(MΩ)	(MΩ)	(MΩ)	(MΩ)	(✓)	(Ω)	(ms)	(ms)

★   Note: Where the installation can be supplied by more than one source, such as a primary source (eg public supply) and a secondary source (eg standby generator), the higher or highest values must be recorded.

TESTED BY

| Signature: | | Position: | | Page 5 of |
| Name: (CAPITALS) | | Date of testing: | |

This form is based on the model shown in Appendix 6 of BS 7671: 2001.
© Copyright Electrical Safety Council (July 2006).

See previous page for Circuit Details

ICM3/9

Circuit impedances

If the circuit is a ring final circuit then R_1, r_n and R_2 must be recorded on some certificates where boxes are provided. On other certificates, $R_1 + R_2$ only may be recorded. This is the measured end to end value of the respective conductor. If the circuit is not a ring final circuit then enter 'n/a' or '/'.

$R_1 + R_2$ This value must be entered for all circuits unless it is not possible to measure. Usually this value is obtained when carrying out the continuity of CPC test.

R_2 Where the measurement of $R_1 + R_2$ is not possible then the end to end resistance of the CPC can be measured using the long lead method. If Table 41C in the *On-Site Guide* is used the CPC impedance is entered here.

Insulation resistance

This is the value of insulation resistance in $m\Omega$ measured between the conductors as identified in the heading. If the installation is measured from the tails, providing the value is greater than the range of the instrument ($>200\,m\Omega$ for example). Then this value can be used for all of the circuits. If the value is less than the range of the instrument, then it would be better to split the installation and measure each circuit individually. A value must be entered; infinity readings are not valid.

Phase to phase readings are for 3 phase circuits.

Polarity

This box is just a tick box to confirm that you have checked polarity. This is normally done when carrying out the continuity of CPC tests.

Live polarity of the incoming supply should be tested at the main switch.

Measured earth fault loop impedance Z_S

This is the measured value of the circuit. The measurement should be taken at the furthest point of the circuit.

It should be compared with the $Z_e + R_1 + R_2$ total and if it is higher, then further investigation should be carried out.

The measured value of $Z_e + R_1 + R_2$ will not include parallel paths if carried out correctly; whereas the measured Z_S will, as this is a live test and all protective conductors must be connected for the test to be carried out safely. Therefore, the measured Z_S should be the same as the $Z_e + R_1 + R_2$ value or even less if parallel paths are present. It should not be higher!

RCD operating times

At IΔn: the actual operating time at the trip rating should be entered here.

At 5 IΔn: this value is only applicable for RCDs used for supplementary protection against direct contact. It should not be measured on any device with an IΔn above 30 mA.

Other

Under the heading of RCD operating times, there is often a column labelled 'other'. This is where the correct mechanical operation of switches, circuit breakers and isolators, etc. are recorded.

Remarks

This area is for the inspector to record anything about the circuit that he/she feels necessary. It may not be a fault but possibly something that may be useful to the next person carrying out an inspection and test on the installation.

Schedule of Items Inspected

This certificate, along with the Schedule of Test Results, forms part of the Electrical Installation Certificate and the Periodic Test Report. Without these schedules the other certificates and reports are invalid.

Completion of this certificate involves the completion of boxes which must be marked using a '✓', 'X' or 'n/a' after the inspection is made, and could be useful as a checklist.

SCHEDULE OF ITEMS INSPECTED † See note below

Methods of protection against electric shock

a. Protection against both direct and indirect contact:

(i) SELV

(ii) Limitation of discharge of energy

b. Protection against direct contact:

(i) Insulation of live parts

(ii) Barriers or enclosures

(iii) Obstacles

(iv) Placing out of reach

(v) PELV

(vi) Presence of RCD for supplementary protection

c. Protection against indirect contact:

(i) EEBAD including:

Presence of earthing conductor

Presence of circuit protective conductors

Presence of main equipotential bonding conductors

Presence of supplementary equipotential bonding conductors

Presence of earthing arrangements for combined protective and functional purposes

Presence of adequate arrangements for alternative source(s), where applicable

Presence of residual current device(s)

(ii) Use of Class II equipment or equivalent insulation

(iii) Non-conducting location: Absence of protective conductors

(iv) Earth-free equipotential bonding: Presence of earth-free equipotential bonding conductors

(v) Electrical separation

Prevention of mutual detrimental influence

a. Proximity of non-electrical services and other influences

b. Segregation of Band I and Band II circuits or Band II insulation used

c. Segregation of safety circuits

Identification

Presence of diagrams, instructions, circuit charts and similar information

Presence of danger notices and other warning notices

Labelling of protective devices, switches and terminals

Identification of conductors

Cables and Conductors

Routing of cables in prescribed zones or within mechanical protection

Connection of conductors

Erection methods

Selection of conductors for current-carrying capacity and voltage drop

Presence of fire barriers, suitable seals and protection against thermal effects

General

Presence and correct location of appropriate devices for isolation and switching

Adequacy of access to switchgear and other equipment

Particular protective measures for special installations and locations

Connection of single-pole devices for protection or switching in phase conductors only

Correct connection of accessories and equipment

Presence of undervoltage protective devices

Choice and setting of protective and monitoring devices (for protection against indirect contact and/or overcurrent)

Selection of equipment and protective measures appropriate to external influences

Selection of appropriate functional switching devices

SCHEDULE OF ITEMS TESTED † See note below

External earth fault loop impedance, Z_e

Installation earth electrode resistance, R_A

Continuity of protective conductors

Continuity of ring final circuit conductors

Insulation resistance between live conductors

Insulation resistance between live conductors and earth

Site applied insulation

Protection by separation of circuits

Protection against direct contact by barrier or enclosure provided during erection

Insulation of non-conducting floors or walls

Polarity

Earth fault loop impedance, Z_s

Operation of residual current devices

Functional testing of assemblies

SCHEDULE OF ADDITIONAL RECORDS* (See attached schedule)

Page No(s)

Note: Additional page(s), must be identified by the page number(s).

† **All boxes must be completed.** '✓' indicates that an inspection or a test was carried out and that the result was **satisfactory**. 'N/A' indicates that an inspection or test was **not applicable** to the particular installation.

Page 3 of

* Where the electrical work to which this certificate relates includes the installation of a fire alarm system and/or an emergency lighting system (or a part of such systems), this electrical safety certificate should be accompanied by the particular certificate(s) for the system(s).

This form is based on the model shown in Appendix 6 of BS 7671: 2001.
© Copyright Electrical Safety Council (July 2006).

ICM3/5

All insulation including inside fuse boards

Electric fence

SELV correctly installed and used

Cables tied to pipes? Hot pipes near cables

OSG p64

Fire alarms and E.M Lighting NOT in same trunking as mains

Check definitions

Limited use. Not domestic

412-05

Earthed LV

412-06

At mains OSG p27+

In circuits

Water Gas Oil etc.

Bathroom

546-02 Extremely rare

Change over sw's?

412-06

Double insulation

Operating theatres

Bonding not domestic

Shaver supply unit

Methods of protection against electric shock

(a) Protection against both direct and indirect contact:
- (i) SELV
- (ii) Limitation of discharge of energy

(b) Protection against direct contact:
- (i) Insulation of live parts
- (ii) Barriers or enclosures
- (iii) Obstacles
- (iv) Placing out of reach
- (v) PELV
- (vi) Presence of RCD for supplementary protection

(c) Protection against indirect contact:
- (i) EEBAD including:
  - Presence of earthing conductor
  - Presence of circuit protective conductors
  - Presence of main equipotential bonding conductors
  - Presence of supplementary equipotential bonding conductors
  - Presence of earthing arrangements for combined protective and functional purposes
  - Presence of adequate arrangements for alternative sources(S), where applicable
  - Presence of residual current devices
- (ii) Use of Class II equipment or equivalent insulation
- (iii) Non-conduction location: Absence of protective conductors
- (iv) Earth-free equipmential bonding: Presence of earth-free equipotential bonding conductors
- (v) Electrical separation

Prevention of mutual deterimental influence
- (a) Proximity of non-electrical services and other influences
- (b) Segregation of band I and band II circuits or band II insulation used
- (c) Segregation of safety circuits ....

Identification
- (a) Presence of diagrams, instruction, circuit chats and similar information
- (b) Presence of danger notices and other warning notices
- (c) Labelling of protective devices, switches and terminal
- (d) Identification of conductors

Cables and conductors
- (a) Routing of cables in prescribed zones or within mechanical protection
- (b) Connection of conductors ....
- (c) Erection methods ....
- (d) Selection of conductors for current-carrying capacity and voltage drop
- (e) Presence of fire barriers, suitables seals and protection against thermal effects

General
- (a) Presence and correct location of appropriate devices for isolation and switching
- (b) Adequacy of access to switchgear and other equipment
- (c) Particular protective measures for special installations and locations
- (d) Connection of single-pole devices for protection or switching in phase conductors only
- (e) Correct connection of accessories and equipment
- (d) Presence of undervoltage protective devices ....
- (g) Choice and setting or protective and monitoring devices for protection against indirect contact and/or overcurrent
- (h) Selection of equipment and protective measures appropriate to external influences ....
- (i) Selection of appropriate functional switching devices

At Mains, Earthing, Bonding. Isolation

CB size, Sw purpose, P/N terminals

Sleeving and cores

OSG p53

Tight no copper showing

Well fitted to BS 7671

Compare with EIC, [orig] or OSG or Calculate

Chapter 42

Is it accessable and safe to work on.

Special location what's required? Part 6

Polarity, VISUAL

As instructions?

Motor Starters

Circuit breakers and fuses correct type or size

Isolator not used as switch

Will the switch do the job?

Notes:
✓ to indicate an inspection has been carried out and the result is satisfactory
X to indicate an inspection has been carried out and the result was unsatisfactory
N/A to indicate the inspection is not applicable

Method of protection against electric shock

Protection against both direct and indirect contact

SELV

If used for protection against direct contact it must not exceed 25 V a.c. or 60 V d.c (Regulations 411-02-01, 412-02-09). Special locations require 12 V a.c. and 30 V d.c.

In domestic installations it is most commonly found supplying extra low voltage lighting, door bells, door entry systems, warden call and security systems.

Commonly found in commercial installations supplying shop window display lighting where halogen luminaires are fed and supported by bare catenary wires.

Warden call and door entry systems should be excluded from the inspection as a simple visual inspection will suffice.

Limitation of discharge of energy

Electric fence controllers use this type of protection which is unlikely to be encountered in a domestic situation. They should be checked for complete segregation (Regulations 411-04-01, 471-03-01, 605-14).

Protection against direct contact

Insulation of live parts

Applies to all electrical installations. This requires the inspection of all live parts for insulation and should be carried out on a sampling basis (Regulations 412-02, 471-04).

Barriers or enclosures

Applies to all electrical installations. All enclosures should comply with Regulation 412-03-01/4; horizontal top surfaces to enclosures IP4X sides, front and bottom IP2X or IPXXB.

Blanks fitted to distribution boards. Access to live parts by use of a key or tool, secondary barriers within enclosures only removable with the use of a key or a tool.

Obstacles

These can be used to prevent unintentional contact with bare live parts. Must be removable without the use of a key or tool but must be secured so as to prevent unintentional removal (Regulations 412-04, 471-06). Not used in domestic installations.

Placing out of reach
Applies to overhead lines. Not likely to be found in domestic installations (Regulations 412-05, 471-07).

PELV
Extra low voltage system, usually with an earthed secondary output. Often used to supply security systems and door entry systems. These systems should not be used in communal systems as there is a risk of imported and exported earth potentials (Regulations 471-14-01/2 restrict the use of PELV).

Presence of RCD for supplementary protection

Where socket outlets could be reasonably expected to supply handheld/portable equipment used outdoors, an RCD with a maximum operating current of 30 mA must be installed (Regulations 412-06-01/2, 471-16).

This method must not be used as the sole means of protection against direct contact.

Protection against indirect contact

Earthed Equipotential Bonding and automatic Disconnection of Supply (EEBADS, Regulation 413-01)

For TN systems, see Regulations 413-02-06/17. For TT systems, see Regulations 413-02-18/20. This will include the following.

Presence of earthing conductor
An installation must have an earthing conductor which must be present before any live testing can be carried out (Regulations 413-02-06, 413-02-18).

This relates to the supply earth which can be an earth electrode (Regulation 542-01).

Presence of circuit protective conductor
Each circuit must have a CPC which must be tested for continuity.

Presence of main equipotential bonding conductor
Regulation 413-02-02. Extraneous parts of the installation must be connected to the Main Earthing Terminal (MET) to comply with Section 547 regulations.

Presence of supplementary equipotential bonding
Regulation 413-02-27. This is required to connect extraneous and exposed conductive parts in situations where there is a greater risk of electric shock.

In a domestic installation the supplementary bonding of bathrooms is a requirement (Regulations 601-04-01/2). This does not exclude the bonding of other areas where a risk is present.

Presence of earthing arrangements for combined and functional purposes
Regulation 607 may apply. Generally for data equipment with protective conductor currents. This current is filtered via capacitors through a functional earth. Equipment metalwork would become live if the earth was lost. This does not apply to domestic installations.

Presence of adequate arrangements for alternative sources where applicable
Normally for generating sets and alternative sources of energy. These sources should be checked to ensure compliance with Regulations 551, 331 and 56. Not normally applicable to domestic installations.

Presence of residual current device
Regulations 413-02-19/20 for TT systems and 413 02-04, 16/17 for TN systems. Also used for fire protection (Regulations 482-02-06 and 605-10-01). Test for correct operation.

Use of Class II equipment or equivalent insulation
Regulations 413-03 and 471-09-03. Double insulated equipment used. Not often found in fixed wiring although possibly found on bell transformers and call systems. Look for markings.

Non-conducting locations
Used in special areas in hospitals and laboratories. Requires specialist knowledge (Regulation 413-04).

Earth free equipotential bonding
Regulation 413-05. Found in electronic repair workshops, specialist knowledge required.

Electrical separation
Regulation 413-06. Used in bathrooms for shaver sockets. The secondary supply is isolated from earth. Does not include SELV.

Prevention of mutual detrimental influence

Proximity of non electrical services and other influences
Regulations 528-02-01/2 and Section 522. Items should be inspected to ensure the electrical system cannot cause harm to non electrical services, and

that the electrical system is unaffected by external influences. Cables tied to pipes or next to central heating pipes for example.

Segregation of band 1 and band II circuits or band II insulation used
Regulations 470-01-02 , 515-01-02, 522-05-03 and Section 528-2/3. Low voltage cables not in same enclosures as extra low voltage cables, such as TV aerials, door bells and telephone cables.

Segregation of safety circuits
Regulations 528-01-04, 563-01-01. Fire alarm and emergency lighting to be segregated from each other and other circuits unless wired in cables with an earthed metal sheath with an insulated covering. This could be Mineral insulated, Firetuff and FP200.

Identification

Presence of diagrams, instructions, circuit charts and similar information
Regulations 514-09-01. Circuit charts, plans, past inspection and test certificates and schedules must be available. For domestic installations, circuit identification would be a minimum requirement on an older installation.

Presence of danger notices and other warning notices
Regulations 541-10-01, 54-11-01, 541-12-01/02, 541-13-01/02. Earth bonding labels, voltage warnings, isolation, harmonization of colours and RCD testing are common.

Labelling of protective devices, switches and terminals
Regulations 514-01-01/2, 514-08-01. Protective devices labelled, conductors in sequence. Switches and isolators marked to identify the item they control if not obvious.

Identification of conductors
Regulations 514-03-01, 514-06-01/2/3/4, 514-07-01/2. Coloured sleeve on switch wires or where phase conductors are not clearly identified.

Cables and conductors

Routing of cables and in prescribed zones or within mechanical protection
Regulations 522-06-01 to 7 or 7.3.1/7.3.2, *On-Site Guide.* Intended for use when completing an Electrical Installation Certificate. Only limited inspection would be possible during a Periodic Inspection Report.

Connection of conductors
Regulations 526-01-01, 526-04-01. Check for tightness and correct use of terminations at a random selection of accessories and all distribution boards.

Erection methods
Regulation Table 4A, Appendix 4 and 522. Correct type of installation to suit environment, standard of workmanship and suitability of fixings.

Selection of conductors for current carrying capacity and voltage drop
Regulation 314-01-03, Chapter 43 and 525, and tabulated values in Appendix 4. Cables to be checked for correct selection with due regard to length, grouping and temperature. Experience is useful here unless design information is available, as generally it would be impractical to carry out volt drop measurements on fully loaded circuits, although this can be calculated using $R_1 + R_2$.

Presence of fire barriers, suitable seals and protection against thermal effects
Regulations, Chapter 42 and Section 527. This is to ensure structural fire barriers are not broken during installation, fire barriers in trunking and ducting where required and intumescent hoods on lighting where installed in fire rated ceilings. Heat from installed equipment is not likely to cause a fire. In particular, check that backless accessories such as wall lights and electrical enclosures installed on surfaces are suitable for surface temperatures and radiated heat of equipment.

General

Presence and correct location of appropriate devices for isolation and switching
Regulations 476, 514-01-01, 537, Chapter 46. Isolators identified where not obvious, local and under the control of the user or, if remote, they must be lockable and identified. Isolation for fans with time control in bathrooms.

Adequacy of access to switchgear and other equipment
Regulations 513-01-01, 526-04-01, 537-04-04, 543-03-03. Doors of enclosures must be removable or able to be fully opened. Access to equipment should not be obstructed. Cooker control switches within 2 metres of cooker and hob (Appendix 8, *On-Site Guide*).

Particular protective measures for special installations and locations
Regulation, Part 6. Bathrooms most common in domestic environment.

Connection of single pole devices for protection and switching in phase conductors only

Regulations 530-01-01/2. Switches in phase and not neutral conductors, generally carried out when continuity of CPC tested along with a visual check.

Correct connection of accessories and equipment

Regulation 712-01-03 and Chapter 526. Check for correct and neat connection, look for excessive exposed conductors, cores of cables not cut out where terminal full.

Presence of undervoltage protective devices

Chapter 45. Motor starters and contactors fitted with correct coils which disconnect on undervoltage and unable to automatically make contacts when correct supply resumed unless physically operated.

Choice and setting of protective and monitoring devices for protection against indirect contact and/or overcurrent

Correct size and type of protective devices/RCD are suitable for use to which they are being put. Monitoring devices not found in domestic situations often found on IT supply systems and some construction sites.

Selection of equipment and protective measures appropriate to external influences

Regulations 512-06-01, 542-01-07, Appendix 5, Chapter 52. Check for suitability of wiring system for the environment and type of use to which it is being put.

Selection of appropriate functional switching devices

Regulations 464 and 537-05. Correct operation of switches/isolators throughout installation.

Periodic Inspection Report

This document is used to record the condition of an installation.
In particular, is it safe to use? At the time of writing, it is not a requirement for the person completing this report to be Part P compliant. It is, however, important that the person carrying out the inspection and test is competent.

The report must also include a Schedule of Test Results and a Schedule of Inspection.

A periodic inspection report is carried out for many reasons, in particular:

    The due date
    Client's/customer's request
    Change of ownership
    Change of use
    Insurance purposes
    To inspect the condition of the existing installation, prior to carrying out
        any alterations or additions

The frequency of the periodic inspection and test is dependent on the type of installation, the environment and the type of use. BS 7671 Wiring Regulations refer to this as the 'Construction, utilization and environment' and this can be found in Appendix 5 of BS 7671.

Guidance note 3 for the inspecting and testing of electrical installations has a table of recommended frequencies for carrying out periodic inspection and tests – the period depends on the type of installation. However, the recommended frequencies are not cast in stone and it is the responsibility of the person carrying out the periodic inspection and test to decide on the period between tests.

This decision should be based on the inspector's experience, what the installation is used for, how often it is used, and the type of environment that surrounds the installation. These things and many others should be taken into account when setting the next test date.

It is important to remember that the date of the first inspection and test is set by the person responsible for the installation design. However, circumstances change which could affect the installation – such as change of use or ownership.

Careful consideration must be given to the installation before the date of the next periodic inspection and test is set.

It is very important that the extent and limitations of the inspection and test is agreed with the person ordering the work before commencing work.

Before the extent and limitation can be agreed, discussion between all parties involved must take place. The client will know why they want the inspection carried out and the person who is carrying out the inspection and test should have the technical knowledge and experience to give the correct guidance.

Past test results, electrical installation, or periodic inspection reports, fuse charts, etc. must be made available to the person carrying out the inspection and test.

If these are not available, then a survey of the installation must be carried out to ensure that the installation is safe to test and to prepare the required paperwork, such as fuse charts.

Whilst carrying out a periodic inspection and test it is not a requirement to take the installation apart. This should be carried out with the minimum of intrusion; disconnection should only be carried out when it is impossible to carry out the required test in any other way. For example, if an insulation resistance test is required on a lighting circuit with fluorescent lighting connected to it.

The simple method would be to open the switch supplying the fluorescent fitting before testing between the live conductors, and close the switch when conducting the test between live conductors and earth. It is not a requirement to disconnect the fitting (see insulation resistance testing, Chapter 4).

Completing the form

As with a Minor Works or an Electrical Installation Certificate, the information required on a Periodic Inspection and Test Report will vary depending on where the report is obtained from.

IPM3

PERIODIC INSPECTION REPORT
FOR AN ELECTRICAL INSTALLATION

Issued in accordance with *British Standard BS 7671- Requirements for Electrical Installations*

A. DETAILS OF THE CLIENT

Client:

Address:

B. PURPOSE OF THE REPORT

This Periodic Inspection Report must be used only for reporting on the condition of an existing installation.

Purpose for which
this report is required:

C. DETAILS OF THE INSTALLATION

Occupier:

Description of premises: Domestic   Commercial   Industrial

Address:

Other:
(Please state)

Estimated age of the electrical installation: years

Postcode:

Evidence of alterations or additions   If yes, estimated age   years

Date of previous inspection:

Electrical Installation Certificate No or previous Periodic Inspection Report No:

Records of installation available:

Records held by:

D. EXTENT OF THE INSTALLATION AND LIMITATIONS OF THE INSPECTION AND TESTING

Extent of the electrical installation covered by this report:

Agreed limitations, if any, on the inspection and testing:

This inspection has been carried out in accordance with BS 7671, as amended. Cables concealed within trunking and conduits, or cables and conduits concealed under floors, in inaccessible roof spaces and generally within the fabric of the building or underground, have not been visually inspected.

E. DECLARATION

I/We, being the person(s) responsible for the inspection and testing of the electrical installation (as indicated by my/our signatures below), particulars of which are described above (see C), having exercised reasonable skill and care when carrying out the inspection and testing, hereby declare that the information in this report, including the observations (see F) and the attached schedules (see H), provides an accurate assessment of the condition of the electrical installation taking into account the stated extent of the installation and the limitations of the inspection and testing (see D). **I/We further declare that in my/our judgement, the said installation was overall in** ✚ _____ **condition (see G) at the time the inspection was carried out, and that it should be further inspected as recommended (see I).**

✚ *(Insert 'a satisfactory' or 'an unsatisfactory', as appropriate)*

INSPECTION, TESTING AND ASSESSMENT BY:

REPORT REVIEWED AND CONFIRMED BY: † *See note below*

Signature:

Signature:

Name:
(CAPITALS)

Name:
(CAPITALS)

Position:

Date:

Date:

† *The completed report should preferably be reviewed by another competent person to confirm that the declared overall condition of the electrical installation is consistent with the inspection and test results, and with the observations and recommendations for action (if any) made in the report.*

Page 1 of ____

This form is based on the model shown in Appendix 6 of BS 7671.
© Copyright The Electrical Safety Council (July 2006)

Please see the 'Notes for Recipients' on the reverse of this page.

IPM3/1

Details of the client Name and address of the person ordering
the work.

Installation address Where is the installation?

Details of the installation *Age.* Often this has to be an estimate based on any
evidence and the inspector's experience. *Alterations.* Have any alterations
been carried out since the last inspection and test? If so how long ago? Are
records available? Who are they kept by?

Purpose of the report Why are you carrying out the inspection and test?

Extent and limitation How much of the installation are you going to inspect?
How much are you going to test? Are there any rooms which you cannot
gain access to, or circuits which must not or cannot be isolated? Generally,
a minimum of 10% of each circuit must be inspected. The extent of the
inspection could be increased if problems/unsafe situations are
discovered.

F. OBSERVATIONS AND RECOMMENDATIONS FOR ACTIONS TO BE TAKEN

Referring to the attached schedules of inspection and test results, and subject to the limitations at D:

There are no items adversely affecting electrical safety.

or

The following observations and recommendations are made.

Item No		Code †
1		

Note: If necessary, continue on additional pages(s), which must be identified by the Periodic Inspection Report date and page number(s).

† Where observations are made, the inspector will have entered one of the following codes against each observation to indicate the action (if any) recommended:-

1. 'requires urgent attention' or
2. 'requires improvement' or
3. 'requires further investigation' or
4. 'does not comply with BS 7671'

Please see the reverse of this page for guidance regarding the recommendations.

Urgent remedial work recommended for Items: **Corrective action(s) recommended for Items:**

G. SUMMARY OF THE INSPECTION

General condition of the installation:

Note: If necessary, continue on additional page(s), which must be identified by the Periodic Inspection Report date and page number(s).

Date(s) of the inspection: Overall assessment of the installation:

(Entry should read either 'Satisfactory' or 'Unsatisfactory')

Page 2 of

Please see the 'Guidance for Recipients on the Recommendation Codes' on the reverse of this page.

IPM3/3

Observations and recommendations

Defects, if any, must be recorded accurately here and a code given to them. Some certificates have codes to indicate the level of attention required. On most certificates the codes will be:

1. **Requires urgent attention** Anything which could compromise the safety of those using the installation should be entered here. This would include lack of earthing, undersized cables, damaged accessories, high Z_s values. It is up to the person carrying out the inspection to make a judgement on this.
2. **Requires improvement** Defects which do not immediately cause the installation to be regarded as unsafe but which could be problematic in the future. This could be: corrosion; old cables such as lighting cables with no CPC terminated in wooden switch boxes; labels missing, etc. Again a judgement must be made by the person carrying out the inspection.
3. **Requires further investigation** This could be anything that the person who is carrying out the inspection and test is concerned about, but which is outside of the agreed extent of the inspection. Possibly a circuit which cannot be traced, or instrument values within the required parameters – anything that might cause concern.
4. **Does not comply with BS 7671** As the requirements of BS 7671 are amended, parts of the installation which would have complied when the installation was new, may now not comply. Examples of this could be: socket outlets that could be used to supply portable equipment outdoors, not protected by an RCD, CPCs which have been sleeved with green sleeving and not green and yellow. Switch returns not identified on older installations.

It must be remembered that it is the responsibility of the person carrying out the inspection to decide on which code to give, the decision should be made using the inspector's experience and common sense.

A true and accurate reflection of the installation must be recorded here. It may require additional pages to explain in detail any observations and recommendations. Do not be influenced by cost or the difficulty in rectifying any defects. The person signing the certificate will be responsible for its content.

Summary of the inspection

This section of the certificate is to detail the overall condition of the installation. It is often easier for the inspector to break the installation into specific areas, for instance:

Any change of use or environment which may have had an affect on the installation
Earthing arrangements
Bonding
Isolation
Age
Safety

Overall assessment

This will either be satisfactory or unsatisfactory. In general terms, if the observation area of the form has any defects other than code 4, the assessment must be unsatisfactory.

H. SCHEDULES AND ADDITIONAL PAGES

Schedule of Items Inspected and Schedules of Items Tested: Page No 4

Additional pages, including additional source(s) data sheets: Page No(s)

Schedule of Circuit Details for the Installation: Page No(s) 5

Schedule of Test Results for the Installation: Page No(s) 6

The pages identified here form an essential part of this report. The report is valid only if accompanied by all the schedules and additional pages identified above.

I. NEXT INSPECTION

I/We recommend that this installation is further inspected and tested after an interval of not more than

(Enter interval in terms of years, months or weeks, as appropriate)

provided that any items at F which have been attributed a Recommendation Code 1 (*requires urgent attention*) are remedied without delay. Items which have been attributed a Recommendation Code 2 or 3 should be actioned as soon as practicable (see F).

J. DETAILS OF ELECTRICAL CONTRACTOR

Trading Title:

Address:

Telephone number:

Fax number:

Postcode:

K. SUPPLY CHARACTERISTICS AND EARTHING ARRANGEMENTS   *Tick boxes and enter details, as appropriate*

✧ **System Type(s)**

TN-S
TN-C-S
TN-C
TT
IT

✧ **Number and Type of Live Conductors**

a.c.

d.c.

1-phase (2 wire)
1-phase (3 wire)
2 pole

2-phase (3 wire)
3-pole

3-phase (3 wire)
3-phase (4 wire)
other

Other   Please state

Nature of Supply Parameters

Nominal voltage(s): $U^{(1)}$ ___ V $U_o^{(1)}$ ___ V

Nominal frequency, $f^{(1)}$ ___ Hz

Prospective fault current, $I_{pf}^{(2)(3)}$ ___ kA

External earth fault loop impedance, $Z_e^{(3)(4)}$ ___ Ω

Number of supplies

Notes:
(1) by enquiry
(2) by enquiry or by measurement
(3) where more than one supply, record the higher or highest values
(4) by measurement

✧ **Characteristics of Primary Supply Overcurrent Protective Device(s)**

BS(EN)

Type

Nominal current rating ___ A

Short-circuit capacity ___ kA

L. PARTICULARS OF INSTALLATION AT THE ORIGIN   *Tick boxes and enter details, as appropriate*

✧ **Means of Earthing**

Supplier's facility:

Installation earth electrode:

Details of Installation Earth Electrode (where applicable)

Type: (eg rod(s), tape etc)

Electrode resistance, R_A: ___ (Ω)

Location:

Method of measurement:

✧ **Main Switch or Circuit-Breaker**
(applicable only where an RCD is suitable and is used as a main circuit-breaker)

Type: BS(EN)

No of Poles

Supply conductors: material

Supply conductors: csa ___ mm²

Voltage rating ___ V

Current rating, I_n ___ A

RCD operating current, $I_{\Delta n}$ ___ mA

RCD operating time (at $I_{\Delta n}$) ___ ms

Maximum Demand (Load): ___ A per phase

Method of Protection against Indirect Contact:

Main Protective Conductors

Earthing conductor	Main equipotential bonding conductors	Bonding of extraneous-conductive-parts (✓)
Conductor material	Conductor material	Water service / Gas service
Conductor csa ___ mm²	Conductor csa ___ mm²	Oil service / Structural steel
Continuity check (✓)	Continuity check (✓)	Lightning protection / Other incoming service(s)

✧ Where a number of sources are available to supply the installation, and where the data given for the primary source may differ from other sources, a separate sheet must be provided which identifies the relevant information relating to each additional source.

Page 3 of ___

This form is based on the model shown in Appendix 6 of BS 7671.
© Copyright The Electrical Safety Council (July 2006)

Please see the 'Notes for Recipients' on the reverse of this page.

IPM3/5

Supply characteristics, earthing and bonding arrangements

Supply characteristics Nominal voltage of the supply.

System type TT, TNS or TNCS.

Nominal frequency Normally 50 Hz.

Prospective fault current Is the highest current that could flow within the installation between live conductors, or live conductors and earth. This should be measured or obtained by enquiry. If it is measured, remember that on a 3 phase system the value between phase and neutral must be doubled.

External earth loop impedance, Z_e This is the external earth fault loop impedance measured between the phase and earthing conductor for the installation.

Characteristics of the supply protective device

BS type Can normally be found printed on the service head.

Nominal current rating Can normally be found printed on the service head.

Short circuit capacity This will depend on the type, but if in doubt reference should be made to Table 7.2A in the *On-Site Guide.*

Main switch or circuit breaker The Type is normally printed on it but reference can be made to Appendix 2 of BS 7671 if required.

Number of poles Does the switch break all live conductors when opened, or is it single pole only?

Supply conductor material and size This refers to the meter tails.

Voltage rating This will usually be printed on the device.

Current rating This will usually be printed on the device.

RCD operating current, IΔn This is the trip rating of the RCD and should only be recorded if the RCD is used as a main switch.

RCD operating time at IΔn Only to be recorded if the RCD is used as a main switch.

Means of earthing

Distributors Facility or earth electrode?

Type If earth electrode.

Electrode resistance Usually measured as Z_e.

Location Where is the earth electrode?

Method of measurement Has an earth fault loop tester or an earth electrode tester been used to carry out the test? To do this test correctly, the earthing conductor should be disconnected to avoid the introduction of parallel paths. This will of course require isolation of the installation; in some instances this may not be practical or possible for various reasons.

If isolation is not possible, the measurement should still be carried out to prove that the installation has an earth. The measured value of Z_e should be equal to or less than any value for Z_e documented on previous test certificates. If the measurement is higher than those recorded before, then further investigation will be required.

The higher measurement could be caused by corrosion, a loose connection or damage.

If the means of earthing is by an earth electrode, the soil conditions may have changed. This would be considered normal providing that the measured value is less than 200Ω and the system is protected by a residual current device.

Main protective conductors

Earthing conductor

Conductor material What is it made of? Unless special precautions are taken in accordance with BS 7671, this should be copper.

Conductor cross-sectional area This must comply with Regulation section 543. If the system is PME then Regulation 547-02-01. In most domestic installations this will require the size to be $16\,\text{mm}^2$. Further information can be found in Tables 10A or 10C of the *On-Site Guide*.

Continuity check This requires a tick only and is usually a visual check, provided that the conductor is visible in its entirety.

Main equipotential bonding conductors

Conductor material What is it made of? Unless special precautions are taken in accordance with BS 7671 this should be copper.

Conductor cross-sectional area This must comply with Regulation 547-02-01. In most domestic installations the required size will be $10\,\text{mm}^2$. Further information is available in Table 10A of the *On-Site Guide*.

Bonding of extraneous conductive parts All services, structural steel, lightning conductors and central heating systems should be equipotential bonded. See Regulation 413-02-02 or Section 4 of the *On-Site Guide*. Normally a tick required.

Safety Certificate for Periodic Inspection

NOTES FOR RECIPIENT

THIS SAFETY CERTIFICATE IS AN IMPORTANT AND VALUABLE DOCUMENT WHICH SHOULD BE RETAINED FOR FUTURE REFERENCE

This safety certificate has been issued to confirm that the electrical installation work to which it relates has been designed, constructed, inspected, tested and verified in accordance with the national standard for the safety of electrical installations, British Standard 7671 (as amended) - *Requirements for Electrical Installations* (formerly known as the IEE Wiring Regulations).

Where, as will often be the case, the installation incorporates a residual current device (RCD), there should be a notice at or near the main switchboard or consumer unit stating that the device should be tested at quarterly intervals. For safety reasons, it is important that you carry out the test regularly.

Also for safety reasons, the complete electrical installation will need to be inspected and tested at appropriate intervals by a competent person. The maximum interval recommended before the next inspection is stated on Page 2 under *Next Inspection*. There should be a notice at or near the main switchboard or consumer unit indicating when the inspection of the installation is next due.

This report is intended for use by electrical contractors not enrolled with NICEIC or by NICEIC Approved Contractors working outside the scope of their enrolment. The certificate consists of at least five numbered pages.

For installations having more than one distribution board or more circuits than can be recorded on pages 4 and 5, one or more additional *Schedules of Circuit Details for the Installation*, and *Schedules of Test Results for the Installation* (pages 6 and 7 onwards) should form part of the certificate.

This certificate is intended to be issued only for a new electrical installation or for new work associated with an alteration or addition to an existing installation. It should not have been issued for the inspection of an existing electrical installation. A 'Periodic Inspection Report' should be issued for such a periodic inspection.

You should have received the certificate marked 'Original' and the electrical contractor should have retained the certificate marked 'Duplicate'.

If you were the person ordering the work, but not the user of the installation, you should pass this certificate, or a full copy of it including these notes, the schedules and additional pages (if any), immediately to the user.

The 'Original' certificate should be retained in a safe place and shown to any person inspecting or undertaking further work on the electrical installation in the future. If you later vacate the property, this certificate will demonstrate to the new user that the electrical installation complied with the requirements of the national electrical safety standard at the time the certificate was issued.

Page 1 of this certificate provides details of the electrical installation, together with the name(s) and signature(s) of the person(s) certifying the three elements of installation work: design, construction and inspection and testing. Page 2 identifies the organisation(s) responsible for the work certified by their representative(s).

Certification for inspection and testing provides an assurance that the electrical installation work has been fully inspected and tested, and that the electrical work has been carried out in accordance with the requirements of BS 7671 (except for any departures sanctioned by the designer and recorded in the appropriate box(es) of the certificate).

If wiring alterations or additions are made to an installation such that wiring colours to two versions of BS 7671 exist, a warning notice should have been affixed at or near the appropriate consumer unit.

Safety Certificate for Electrical Installation

NOTES FOR RECIPIENTS

THIS SAFETY CERTIFICATE IS AN IMPORTANT AND VALUABLE DOCUMENT
WHICH SHOULD BE RETAINED FOR FUTURE REFERENCE

The purpose of periodic inspection is to determine, so far as is reasonably practicable, whether an electrical installation is in a satisfactory condition for continued service. This report provides an assessment of the condition of the electrical installation identified overleaf at the time it was inspected, taking into account the stated extent of the installation and the limitations of the inspection and testing.

The report has been issued in accordance with the national standard for the safety of electrical installations, British Standard 7671 (as amended) - *Requirements for Electrical Installations* (formerly known as the IEE Wiring Regulations).

Where the installation incorporates a residual current device (RCD), there should be a notice at or near the main switchboard or consumer unit stating that the device should be tested at quarterly intervals. For safety reasons, it is important that you carry out the test regularly.

Also for safety reasons, the electrical installation will need to be re-inspected at appropriate intervals by a competent person. The recommended maximum time interval to the next inspection is stated on page 3 in Section I (*Next Inspection*). There should be a notice at or near the main switchboard or consumer unit indicating when the next inspection of the installation is due.

The report consists of at least six numbered pages. The report is invalid if any of the pages identified in Section H are missing.

For installations having more than one distribution board or more circuits than can be recorded on Pages 5 and 6, one or more additional *Schedules of Circuit Details for the Installation*, and *Schedules of Test Results for the Installation* (pages 7 and 8 onwards) should form part of the report.

This report is intended to be issued only for the purpose of reporting on the condition of an existing electrical installation. The report should identify, so far as is reasonably practicable and having regard to the extent and limitations recorded in Section D, any damage, deterioration, defects, dangerous conditions and any non-compliances with the requirements of the national standard for the safety of electrical installations which may give rise to danger. It should be noted that the greater the limitations applying to a report, the less its value.

The report should not have been issued to certify that a new electrical installation complies with the requirements of the national safety standard. An 'Electrical Installation Certificate' or a 'Domestic Electrical Installation Certificate' (where appropriate) should be issued for the certification of a new installation.

You should have received the report marked 'Original' and the electrical contractor should have retained the report marked 'Duplicate'.

If you were the person ordering the work, but not the user of the installation, you should pass this report, or a full copy of it including these notes, the schedules and additional pages (if any), immediately to the user.

The 'Original' report form should be retained in a safe place and shown to any person inspecting or undertaking further work on the electrical installation in the future. If you later vacate the property, this report will provide the new user with an assessment of the condition of the electrical installation at the time the periodic inspection was carried out.

Section D addresses the extent and limitations of the report by providing boxes for the *Extent of the electrical installation covered by this report* and the *Agreed limitations, if any, on the inspection and testing*. Information

NOTES FOR RECIPIENT
(continued from the reverse of page 1)

Where responsibility for the *design*, the *construction* and the *inspection and testing* of the electrical work is divided between the electrical contractor and one or more other bodies, the division of responsibility should have been established and agreed before commencement of the work. In such a case, the absence of certification for the *construction*, or the *inspection and testing* elements of the work would render the certificate invalid. If the *design* section of the certificate has not been completed, you should question why those responsible for the design have not certified that this important element of the work is in accordance with the national electrical safety standard.

All unshaded boxes should have been completed either by insertion of the relevant details or by entering 'N/A', meaning 'Not Applicable', where appropriate.

Where the electrical work to which this certificate relates includes the installation of a fire alarm system and/or an emergency lighting system (or a part of such systems) in accordance with British Standards BS 5839 and BS 5266 respectively, this electrical safety certificate should be accompanied by a separate certificate or certificates as prescribed by those standards.

Where the installation can be supplied by more than one source, such as the public supply and a standby generator, the number of sources should have been recorded in the box entitled Number of Supplies, under the general heading *Supply Characteristics and Earthing Arrangements* on page 2 of the certificate, and the *Schedule of Test Results* compiled accordingly. Where a number of sources are available to supply the installation, and where the data given for the primary source may differ from other sources, an additional page should have been provided which gives the relevant information relating to each additional source, and to the associated earthing arrangements and main switchgear.

GUIDANCE FOR RECIPIENTS ON THE RECOMMENDATION CODES

Only one Recommendation Code should have been given for each recorded observation.

Recommendation Code 1

Where an observation has been given a Recommendation Code 1 (requires urgent attention), the safety of th using the installation may be at risk.

The person responsible for the maintenance of the installation is advised to take action without delay to rem the observed deficiency in the installation, or to take other appropriate action (such as switching off and isola the affected part(s) of the installation) to remove the potential danger. The electrical contractor issuing this re will be able to provide further advice.

It is important to note that the recommendation given at Section I *Next Inspection* of this report for the maxi interval until the next inspection, is conditional upon all items which have been given a Recommenda Code 1 being remedied without delay.

Recommendation Code 2

Recommendation Code 2 (requires improvement) indicates that, whilst the safety of those using the installa may not be at immediate risk, remedial action should be taken as soon as possible to improve the safety o installation to the level provided by the national standard for the safety of electrical installations, BS 7671. electrical contractor issuing this report will be able to provide further advice.

Items which have been attributed Recommendation Code 2 should be remedied as soon as possible Section F).

Recommendation Code 3

Where an observation has been given a Recommendation Code 3 (requires further investigation), the inspe has revealed an apparent deficiency which could not, due to the extent or limitations of this inspection, be identified. Items which have been attributed Recommendation Code 3 should be investigated as soon as pos (see Section F).

The person responsible for the maintenance of the installation is advised to arrange for the further examinati the installation to determine the nature and extent of the apparent deficiency.

Recommendation Code 4

Recommendation Code 4 [does not comply with BS 7671 (as amended)] will have been given to obse non-compliance(s) with the **current** safety standard which do not warrant one of the other Recommend Codes. It is not intended to imply that the electrical installation inspected is unsafe, but careful consider should be given to the benefits of improving these aspects of the installation. The electrical contractor issuing report will be able to provide further advice.

5

Safety in electrical testing

Correct selection of protective devices

(While protective devices are mentioned throughout this book, this chapter brings all of the information together for reference)

When carrying out an inspection and test on any electrical installation it is important to ensure that the correct size and type of device has been installed.

To do this we must have a good knowledge of the selection of protective devices and the type of circuits that they are protecting.

Why are they installed?

Protective devices are installed to protect the cable of the circuit from damage – this could be caused by <u>overload, overcurrent and fault current</u>.

The definition for overload given in Part 2 of BS 7671:2004 is: *Overcurrent occurring in a circuit which is electrically sound*. This is when the circuit is installed correctly and the equipment connected to it is drawing too much current. For instance:

An electric motor connected to the circuit is used on too heavy a load, leading to an overload of the circuit. Provided that the correct size of protective device was installed, the device will operate and interrupt the supply preventing the cable from overloading.

If additional luminaries were installed on an existing circuit which was already fully loaded, the protective device should operate and protect the cable of the circuit.

Overcurrent is a current flow in a circuit which is greater than the rated current carrying capacity of the cables. This would normally be due to a fault on the circuit or incorrect cable selection. For example:

> If a 20 amp cable protected by a 32 amp circuit breaker was loaded by 25 A, the cable would overheat and the device would continue to allow current to flow – this could damage the cable.

A fault current is a current which is flowing in a circuit due to a fault. For example:

> A nail is driven through a cable causing an earth fault or a short circuit fault. This would cause a very high current to flow through the circuit, which must be interrupted before the conductors reach a temperature that could damage the insulation or even the conductors.

So what are we looking for with regard to protective devices during an inspection?

What type of device is it? Is it a fuse or circuit breaker?

A fuse has an element which melts when too much current is passed through it, whether by overload or fault current.

Fuses in common use are:

BS 3036 semi-rewirable fuse
BS 88 cartridge fuse
BS 1361 cartridge fuse

A circuit breaker is really two devices in one unit. The overload part of the device is a thermal bi metal strip, which heats up when a current of a higher value than the nominal current rating (I_n) of the device passes through it.

Also incorporated within the device is a magnetic trip, which operates and causes the device to trip when a fault current flows through it. For the device to operate correctly it must operate within 0.1 seconds. The current which has to flow to operate the device in the required time has the symbol (I_a).

Circuit breakers in common use are:

☞ BS 3871 types 1, 2 and 3
☞ BS EN types B, C and D (*A is not used, this is to avoid confusion with Amps*)

Is the device being used for protection against indirect contact?

In most instances this will be the case.

What type of circuit is the device protecting, is it supplying fixed equipment only, or could it supply handheld equipment?

If the circuit supplies fixed equipment only, the device must operate on fault current within 5 seconds. If it supplies socket outlets it must operate within 0.4 seconds (*see BS 7671, Regulations 413-02-08 and 413-02-09*).

When using circuit breakers to BS 3871 and BS EN 60898 these times can be disregarded. Providing the correct Z_S values are met, they will operate in 0.1 seconds or less.

If it is a circuit breaker is it the correct type?

Table 7.2b of the *On-Site Guide* provides a good reference for this.

Types 1 and B should be used on circuits having only resistive loads (*have you ever plugged in your 110 v site transformer and found that it operated the circuit breaker? If you have it will be because it was a type 1 or B*).

Types 2, C and 3 should be used for inductive loads such as fluorescent lighting, small electric motors and other circuits, where surges could occur.

Types 4 and D should be used on circuits supplying large transformers or any circuits where high inrush currents could occur.

Will the device be able to safely interrupt the prospective fault current which could flow in the event of a fault?

Table 7.2A or the manufacturer's literature will provide information on the rated short circuit capacity of protective devices.

Is the device correctly coordinated with the load and the cable?

Correct coordination is:

> Current carrying capacity of the cable under its installed conditions must be equal to or greater than the rated current of the protective device (I_z).

The rated current carrying capacity of the protective device (I_n) must be equal to or greater than the design current of the load (I_b).

In short, $I_z > I_n > I_b$ (*Appendix 4. item 4 BS 7671 or Appendix 6 On-Site Guide*).

Additional information regarding circuit breakers

Overload current
The symbol for the current required to cause a protective device to operate within the required time on overload is (I_2).

Circuit breakers with nominal ratings up to 60 amps must operate within 1 hour at $1.45 \times$ their nominal rating (I_n).

Circuit breakers with nominal ratings above 60 amps must operate within 2 hours at $1.45 \times$ its rating (I_n).

At 2.55 times the nominal rating (I_n), circuit breakers up to 32 amps must operate within 1 minute; and circuit breakers above 32 amps, must operate within 2 minutes.

They must <u>not</u> trip within 1 hour at up to 1.13 their nominal rating (I_n).

Maximum earth fault loop impedance values (Z_S) for circuit breakers
These values can be found in Table 41B2 in BS 7671, they can also be found in Sections 604 and 605 of BS 7671.

Because these devices are required to operate within 0.1 of a second, they will satisfy the requirements of BS 7671 with regard to disconnection times in all areas. Therefore, the Z_S values for these devices are the same wherever they are to be used (this only applies to circuit breakers) even in special locations where the disconnection time must be 0.2 seconds.

Calculation of the Maximum Z_S of circuit breakers
It is often useful to be able to calculate the maximum Z_S value for circuit breakers without the use of tables. This is quite a simple process for BS 3871 and BS EN 60898 devices. Let's use a 20A BS EN 60898 device as an example:

Table 7.2b shows that a type B device must operate within a window of 3 to 5 times its rating. As electricians we always look at the worst case scenario. Therefore, we must assume that the device will not operate until a current equal to 5 times its rating flows through it (I_a). For a 20 A type B device this will be $5 \times 20\ A = 100\ A$.

If we use a supply voltage of 240 volts – which is the assumed open circuit voltage (U_{OC}) of the supply (Appendix 3, BS 7671) – Ohm's law can be used to calculate the maximum Z_S.

$$Z_S = \frac{U_{OC}}{I_a}$$

$$Z_S = \frac{240}{100} = 2.4\ \Omega$$

And to check this, we look in Table 41B2 of BS 7671, we will see that the value Z_S for a 20 A type B device is $2.4\ \Omega$.

Now let's use the same procedure for a 20 A type C device. Table 7.2b shows us that a type C device must operate at a maximum of 10 times its rating (I_n).

$$10 \times 20 = 200\ A$$

$$\frac{240}{200} = 1.2\ \Omega$$

If we check again in Table 41B2, we will see that the maximum Z_S for a 20 A type C device is $1.2\ \Omega$.

A type C circuit breaker with a nominal operating current (I_n) must operate at a maximum of 20 times its rating:

$$20 \times 20 = 400\ A$$

$$\frac{240}{400} = 0.6\ \Omega$$

Again, if we check in Table 41B2 we will see that the Z_S value is $0.6\ \Omega$.

We can see that the maximum Z_S values for a type C are 50% of the Z_S value of a type B device, and that the Z_S value for a type D are 50% of the Z_S value of a type C device.

Maximum earth loop impedance (Z_S) for fuses

Maximum earth loop impedance (Z_S) for fuses

Fuses have to operate at 0.4 or 5 seconds depending on the type of circuit which they are protecting. To find the maximum permissible Z_S value for a fuse, the current curves in Appendix 3 of BS 7671 will have to be looked at.

Example

Find the maximum permissible Z_S for a BS 1361 fuse with a rating of 20 amps with a required disconnection time of 0.4 seconds.

Look in Appendix 3, Figure 3.1. The left-hand side of the grid represents the disconnection time.

From the bottom left-hand corner, follow the line upwards until the horizontal line representing 0.4 seconds is found. Now follow the horizontal line across to the right until the bold line for a 20 amp fuse is found, from where the horizontal line touches the bold line move vertically down the page until you meet the bottom line of the grid. The bottom line represents the automatic operating current for the fuse. It can be seen that the current required is around 130 amps.

The table in the right-hand top corner of the page will show this value to be 135 amps.

The calculation is:

$$Z_S = \frac{U_{OC}}{I_a}$$

$$Z_S = \frac{240}{135}$$

$$Z_S = 1.777 \ \Omega$$

Rounded up, this is 1.77 Ω. As a check, look in BS 7671, part 4, chapter 1, Table 41B1 and you will see that the Z_S for a BS 1361 20 amp fuse is 1.77Ω.

Do not forget to correct this value for the conductor operating temperature and ambient temperature, if required.

This calculation can be used for any type of protective device. But remember that the disconnection time for a circuit breaker will always need to be 0.1 seconds, and that for some special locations it will be 0.2 seconds.

Comparing maximum Z_S and measured Z_S

Unfortunately, we cannot compare this value directly to any measured Z_S values that we have. This is because the values given in BS 7671 for Z_S are for when the circuit conductors are at their operating temperature (generally 70°C).

However, we can use a simple calculation which is called the rule-of-thumb (Section 2.7.14 of *Guidance note 3*). This calculation will allow us to compare our measured values with the values from BS 7671.

The values given in the Table below are the worst case values. In these types of calculations we must always use the worst case values to ensure a safe installation.

From this table below, it can be seen that types D, 3 and 4 will have very low maximum permitted Z_S values. This will often result in the use of an RCD.

Circuit breaker type		Worst case tripping current	Typical uses
BS EN 60898	B	5 times its rating	General purpose with very small surge currents (i.e. small amounts needed for fluorescent lighting). Mainly domestic
	C	10 times its rating	Inductive loads. Generally commercial or industrial where higher switching surges would be found (i.e. large amounts of fluorescent lighting or motor starting)
	D	20 times its rating	Only to be used where equipment with very high inrush currents would be found
BS 3871	1	4 times its rating	As for type B
	2	7 times its rating	As for type C
	3	10 times its rating	As for type C but slightly higher inrush currents
	4	50 times its rating	As for type D

Example

Let's assume that we have a circuit protected by a 32A BS EN 60898 type B device. The measured value of Z_S is $0.98\,\Omega$.

Following the procedure described previously:

$$5 \times 32 = 160 \ A$$

$$\frac{240}{160} = 1.5 \ \Omega$$

The maximum Z_S at 70°C for the circuit is 1.5 Ω.

To find $\frac{3}{4}$ of 1.5 we can multiply it by 0.75.

$$1.5 \times 0.75 = 1.125\,\Omega$$

So, 1.125 Ω now becomes our maximum value, and we can compare our measured value directly to it without having to consider the ambient temperature or the conductor operating temperature. Our measured value must be less than the corrected maximum value; in this case it is, and the 32 amp type B device would be safe to use.

> This type of calculation must be understood by any student studying for the City and Guilds 2391 inspecting and testing course

Test equipment

It is important that the test equipment you choose is suitable for your needs. Some electricians prefer to use individual items of equipment for each test; others like to use multi-function instruments.

Any test instruments used for testing in areas such as petrol filling stations, or areas where there are banks of storage batteries, etc. In fact, anywhere there is a risk of explosion, must be intrinsically safe for the purpose.

Most electricians are aware that electrical test instruments must comply with BS standards. However, students studying for Part P and City and Guilds 2391 exams, an understanding of the basic operational requirement for the most common types of test instrument is very important.

Whichever instrument you choose it must be suitable for the use to which it is to be put and be manufactured to the required British standard. It is also vital that you fully understand how to operate it before you start testing.

Instruments required

Low resistance ohm meter

Used to measure the resistance and verify the continuity of conductors. This instrument must produce a test voltage of between 4 and 24 volts and a current of not less than 200 mA. The range required is 0.02 to 2 ohms although most modern instruments are self ranging and will measure higher values if required.

Insulation resistance tester

Used to measure the insulation resistance between live conductors and live conductors and earth. This test is a pressure test of conductor insulation. The instrument must deliver a current of 1 mA on a resistance of 0.5 MΩ. At 250 volts d.c. for extra low voltage circuits, 500 volts d.c. for low voltage circuits up to 500 volts a.c. and 1000 volts d.c. for circuits between 500 and 1000 volts d.c.

This instrument is sometimes called a high resistance tester as it measures values in megohms.

Earth fault loop impedance tester

This instrument allows a current of up to 25 amps to flow around the earth fault loop path. It measures the current flow and by doing so can calculate the resistance of the earth fault loop path. The values given are in ohms.

As a current of 25 amps will trip RCDs and some smaller circuit breakers, it is useful to have an instrument that can carry out low current testing where required. Use of this type of instrument will avoid the tripping of devices during testing.

Prospective short circuit current test instrument

This instrument measures the current that would flow between live conductors in the event of a short circuit. It is usually incorporated in the earth loop impedance tester and normally gives a value in kA. Some instruments give the value in ohms which then needs to be converted to amps by using Ohm's law (*use 240 volts*).

Most instruments will measure the value between phase and neutral and not between phases; to find the value between phases it is simply a matter of doubling the phase to neutral value.

Earth electrode resistance tester

This is normally a battery operated 3 or 4 terminal instrument with a current and potential spike. The values given are in ohms and the instrument instructions should be fully understood before using it. The instrument would be used where low and very accurate earth electrode resistance values are required such as for generators or transformers.

Residual current device tester

This instrument measures the tripping times of RCDs in seconds or milliseconds.

Phase rotation

This instrument is used to ensure the correct phase rotation of 3 phase supplies.

Calibration of test instruments

To carry out any kind of test properly your instruments have to be accurate. If they were not then the whole point of carrying out the test would be lost. While it is not a requirement to have instruments calibrated on an annual basis, a record must be kept to show that the instruments are regularly checked for accuracy.

Instrument accuracy can be tested using various methods. For an earth loop impedance tester all that is required is a dedicated socket outlet. Use your earth fault loop impedance instrument to measure the value of the socket outlet. This value can then be used as a reference to test the accuracy of the instrument at a later date. You can also test any other earth fault loop instrument on the dedicated outlet to check its accuracy. The loop impedance values of the socket outlet should not change.

For an insulation resistance tester, or a low resistance ohm meter, the accuracy can be checked quite simply by using various values of resistors. The instruments could even be checked against values given by another instrument. If the values given are not the same when testing against another instrument, this will indicate that one of the instruments is inaccurate and further investigation using resistors should be undertaken.

An RCD test instrument accuracy is a little more difficult to check and often the best way is to check it against another instrument. However, if you do check it in this way do not expect exactly the same values as the trip time could be slightly different each time you test it due to the instrument increasing slightly in temperature.

It is also possible to purchase a calibration instrument which will check the accuracy of all electrical test instruments. When it is found that an instrument is not accurate, then it must be returned to the manufacturer or specialist for re-calibration. This is not a job that can normally be carried out by the owner of the instrument.

Testing for continuity

Calibration instrument

Testing for insulation

Calibration instrument

If, for any reason, your instrument does require re-calibration, it should be returned to the instrument manufacturer or a calibration specialist.

Recordkeeping for accuracy testing of the instrument is quite a simple but important process. A record showing the instrument model, serial number and the date of the test along with the recorded values is all that is required and will satisfy most regulatory bodies. Records can be kept in a ledger, on a computer. Alternatively, calibration registers can be purchased to make life a little easier.

Electric shock

An electric shock is caused by current flowing through a body. A very small amount – between 50 and 80 mA is considered to be lethal to most human beings, although this would of course depend on the person's health and other circumstances. In livestock the lethal current would be considerably less.

The electrical regulations are set out to provide for the safety of persons and livestock. An electric shock is one risk of injury; others are:

* Excessive temperatures likely to cause burns, fire and other injurious effects.
* Mechanical movement of electrically actuated equipment, in so far as such injury is intended to be prevented by electrical emergency switching or by switching for mechanical maintenance of non-electrical parts of such equipment.
* Explosion.

This can be found in Chapter 13 of BS 7671 (Regulation 130-01-01).

Regulation 130-02-01 tells us that persons and livestock shall be protected so far as is reasonably practical against dangers that may arise from contact with live parts of the installation. This protection can be achieved by one of the following methods.

For direct or indirect contact:

* Preventing current passing through the body of any person or livestock.
* Limiting the current which can pass through a body to a value lower than the shock current.

KEWTECH

TEST EQUIPMENT RECORD FASTCHECK 1000

COMPANY:
INSTRUMENT USED:
SERIAL NO:
NEXT TEST DATE:
Page ___ of ___ Cert No.

TESTED BY	MODEL	SERIAL No.	DATE	INSULATION V	INSULATION RANGE MΩ	INSULATION READING MΩ	CONTINUITY RANGE Ω	CONTINUITY READING Ω	LOOP LOCAL LOOP Ω	LOOP +0.5Ω	RCD 50ms 15mA	RCD 50ms 30mA	RCD 40ms 150mA	Dmm ACV 230v+/-10%	Dmm DCV 110v+/-10%	RESULTS PASS	RESULTS FAIL x
Mr. Smith	KT35	12345678	10.09.04	250	1.9MΩ	1.9MΩ	1.9Ω	1.8Ω								✓	
				500	9.9MΩ	9.8MΩ	9.9Ω	9.8Ω								✓	
				1000	99MΩ	99MΩ	99Ω	99Ω								✓	
Mr. Smith	KTA2	23456789	10.09.04						0.50	1.0						✓	
Mr. Smith	KT56	92345678	10.09.04								50ms	49ms	40ms				x
Mr. Smith	RCD1	85432	15.9.04								40ms	41ms	29ms				x
Mr. Smith	LOOP1	145678	15.9.04						0.50	1.5							x
Mr. Smith	MFT2	876321	15.9.04	250	1.9MΩ	1.9MΩ	1.9Ω	1.8Ω	0.50	0.99	49ms	49ms	39ms			✓	
				500	9.9MΩ	9.8MΩ	9.9Ω	9.8Ω									
				1000	99MΩ	99MΩ	99Ω	99Ω									
Mr. Smith	Dmm3	546245	15.9.04											233V	111V	✓	
Mr. Smith	Ins 2	678910	16.09.04	500	1.9MΩ	1.8MΩ	1.9Ω	1.9Ω								✓	
				500	9.9MΩ	10MΩ	9.9Ω	9.9Ω									
				500	99MΩ	100MΩ	99Ω	100Ω									

For indirect contact:

- Automatic disconnection of the supply in a determined time on the occurrence of a fault likely to cause a current to flow through a body in contact with exposed conductive parts, where the value of that current is equal to or greater than the shock current.

The regulations quoted mention **Direct** and **Indirect** contact.

> Direct contact is electric shock received by touching a known live part which is intended to be live.
>
> Indirect contact is electric shock received when touching exposed conductive parts made live due to a fault.

Protection can be achieved by various methods which can be found in Chapters 41 and 47 of BS 7671. The most common methods used for protection against direct contact within electrical installations are: the use of insulation, and enclosures.

Protection from electric shock from indirect contact can be by many methods. The most common method used within a normal electrical installation is by the use of Earthed Equipotential Bonding and the Automatic Disconnection of the supply (EEBADs). Class 2 equipment (double insulated) and electrical separation (shaver socket) are also very common methods.

In a single phase system, current flow is achieved by creating a difference in potential.

If we were to fill a tank with water and raise the tank a metre or so, then connect a pipe with a tap on one end of it to the tank, when we open the tap the water will flow from the tank to the open end of the pipe. This is because there is no pressure outside of the pipe – the higher we raise the tank the greater the pressure of water and therefore the greater the flow of water.

Current flow is very similar to this. If we think of voltage as pressure, then to get current to flow we have to find a way of creating a difference in pressure. This pressure in an electrical circuit is called **potential difference** and it is achieved in a single phase system by pegging the star point of the supply transformer to earth. The potential of earth is known to be at 0 volts.

If we place a load between a known voltage and earth, the current will flow from the higher voltage through the load to earth. If we increase the voltage, then more current will flow, just as more water would flow if we increased the height of the water tank.

The problem we have with electricity is that if we use our body to provide the current with a path to earth it will use it, and possibly electrocute us at the same time.

Current will not flow unless it has somewhere to flow to – that is, from a high pressure to a lower pressure, possibly zero volts but not always. It is also possible in some instances to get different voltages in an installation, particularly during a fault where volt drops may occur due to loose connections, high resistance joints and different sizes of conductors. We must also remember that during a fault it will not only be the conductors that are live, but any metalwork connected to the earthing and bonding system, either directly or indirectly. It is highly likely that a shock by indirect contact could be received between pipe work at different voltages.

In any installation, protection must be in place to prevent electric shock. The protection we use against direct contact is self-explanatory and we can only prevent unintentional touching of live parts – if a person is intent on touching a live conductor, we can only make it difficult for them, not impossible.

Protection against indirect contact is a different problem altogether and we can achieve it by different methods. Firstly, if there is a fault to earth all of the metal work connected to the earthing system, whether directly or indirectly, would become live. In the first instance we need to ensure that enough current will flow through the protective device to earth to operate the protective device very quickly. This is achieved by selecting the correct type of protective device, and ensuring that the earth fault loop path has a low enough impedance to allow enough current to flow and operate the device in the required time. On its own this is not enough and that is where the equipotential and supplementary bonding is used. The basic principle is that, if one piece of metal work becomes live, any other parts that could introduce a potential (voltage) difference also become live at the same potential. If everything within the building is at the same potential, current cannot possibly flow from one part to another via a person or livestock.

Testing transformers

It is a requirement to test isolation and SELV transformers to ensure the user's safety. It is also useful to be able to test them to ensure that they are working correctly.

Step up or down double wound transformer

Use a low reading ohm meter to test between to primary (*cable that connects to the main supply*) side. The resistance should be quite high – this will of course depend on the size of the transformer. It may be that the resistance is so high that a multi meter set on its highest resistance value will have to be used. If this is the case, then set the instrument to the highest value possible and turn it down until a reading is given. If the winding is open circuit then the transformer is faulty. Repeat this test on the secondary winding.

Now join the ends of the primary winding together and join the ends of the secondary winding together. Use an insulation resistance meter set on 500 V d.c. to test between the joined ends. Then test between the joined ends and earth. The maximum insulation value permissible in both cases is 0.5 MΩ. If the resistance is less, then the transformer is faulty.

Isolation transformer

Carry out the test in the same manner as the double wound transformer and the values for insulation resistance are the same, 0.5 MΩ.

Separated Extra Low Voltage transformers (SELV)

These transformers are tested using the same procedure as for the step up or down transformer. The insulation resistance test values are different for this test. If the SELV circuits from the secondary side of the transformer are being tested, then the test voltage must be 250 V d.c. and the maximum resistance value is 0.25 MΩ – although this would be considered a very low value and any value below 5 MΩ must be investigated.

The test voltage is increased to 500 V d.c. for a test between the actual transformer windings. The minimum insulation resistance value is 0.5 MΩ although any value below 5 Ω must be investigated.

Testing a 3 phase induction motor

There are many types of 3 phase motors but by far the most common is the induction motor. It is quite useful to be able to test them for serviceability.

Before carrying out electrical tests it is a good idea to ensure that the rotor turns freely. This may involve disconnecting any mechanical loads. The rotor should rotate easily and you should not be able to hear any rumbling from the motor bearings. Next, if the motor has a fan on the outside of it, check that it is clear of any debris which may have been sucked in to it. Also check that any air vents into the motor are not blocked.

Generally, if the motor windings are burnt out there will be an unmistakable smell of burnt varnish. However, it is still a good idea to test the windings as the smell could be from the motor being overloaded. Three phase motors are made up of three separate windings – in the terminal box there will be six terminals as each motor winding will have two ends. The ends of the motor windings will usually be identified as W1, W2; U1, U2; or V1, V2. The first part of the test is carried out using a low resistance ohm meter. Test each winding end to end (W1 to W2, U1 to U2 and V1 to V2). The resistance of each winding should be approximately the same and the resistance value will depend on the size of the motor. If the resistance values are different, then the motor will not be electrically balanced and it should be sent for rewinding. If resistance values are the same, then the next test is carried out using an insulation resistance tester. Join W1 and W2 together, U1 and U2 together and V1 and V2 together. Carry out an insulation resistance test between the joined ends, i.e. W to U then W to V and then between U and V. Then repeat the test between joined ends and the case, or the earthing terminal of the motor (*these tests can be in any order to suit you*). Providing the insulation resistance is $2\,M\Omega$ or greater then the motor is fine. If the insulation resistance is above $0.5\,M\Omega$ this could be due to dampness and it is often a good idea to run the motor for a while before carrying out the insulation test again as the motor may dry out with use.

To reconnect the motor windings in star, join W2, U2 and V2 together and connect the 3 phase motor supply to W1, U1 and V1. If the motor rotates in the wrong direction, swap two of the phases of the motor supply.

To reconnect the motor windings in delta, join W1 to U2, U1 to V2 and V1 to W2 and then connect the 3 phase motor supply one to each of the joined ends. If the motor rotates in the wrong direction, swap two phases of the motor supply.

Appendix A

IP codes

Ingress protection

In BS 7671 Wiring Regulations the definition of an enclosure is 'A Part providing protection of equipment against certain external influences and in any direction against direct contact'.

To ensure that we use the correct protection to suit the environment where the enclosure is installed, codes are used. These codes are called IP codes. IP stands for ingress protection and is an international classification system for the sealing of electrical enclosures or equipment.

The system uses the letters IP followed by two or three digits. The first digit indicates the degree of protection required for the intrusion of foreign bodies such as dust, tools and fingers.

The second digit provides an indication of the degree of protection required against the ingress of moisture.

If a third digit is used, a letter would indicate the level of protection against access to hazardous parts by persons; a number would indicate the level of protection against impact.

Where an 'X' is used, it is to show that nothing is specified. For example, if a piece of equipment is rated at IPX8, it would require protection to allow it to be submersed in water. Clearly if a piece of equipment can be submersed safely, then dust will not be able to get in to it and no protection against the ingress of dust would be required.

Table of IP ratings

Dust and foreign bodies	Level of Protection	Moisture	Level of Protection
0	No special protection	0	No special protection
1	50 mm	1	Dripping water
2	12.5 mm diameter and 80 mm long (finger)	2	Dripping water when tilted at 15°
3	2.5 mm	3	Rain proof
4	1 mm	4	Splash proof
5	Limited dust	5	Sprayed from any angle (jet proof)
6	Dust tight	6	Heavy seas and powerful jets
		7	Immersion up to 1 M
		8	Submersion 1 M+

Third digit, usually a letter

A	The back of a hand or 50 mm sphere
B	Standard finger 80 mm long
C	Tool 2.5 mm diameter, 100 mm long, must not contact hazardous areas
B	Wire 1 mm diameter, 100 mm long, must not contact hazardous areas

The third number for impact is not used in BS 7671 and is not included in this book.

Appendix

B

Exercises

Exercise 1

Mrs F. G. Grant who lives in Bath, Somerset, has asked you to install a new 32 A/230 V cooker circuit installed in the Kitchen.

The cooker control unit incorporates a 13 amp socket outlet. The existing installation is in good condition and complies with BS 7671, 2001 amended to 2004. A recent Periodic Test Report, Schedules of Inspection and Test Results are available.

The wiring for the new circuit is $6\,\text{mm}^2$ thermoplastic (pvc) flat twin with CPC cable. The circuit protection is by a type B32A RCBO to BS EN 61009 which has a maximum Z_S of 1.5 ohms and an Icn rating of 6 kA. This protective device is housed in a consumer unit, which is situated in the integral garage of the house. The main switch is BS EN 60439, 100 A 230 V.

The ambient temperature is 20°C, $R_2 + R_2$ value for this cable is 10.49 milli-ohms per metre, the circuit is 32 metres long.

The supply is TN-S 230 V 50 Hz, the main fuse is 100A BS 1361, and the measured values of PFC and Z_e are 800 amp and 0.24 ohms, respectively. The supply tails are $25\,\text{mm}^2$ copper; the earthing conductor is $16\,\text{mm}^2$ copper; and the main equipotential bonding to gas and water is $10\,\text{mm}^2$ copper. Maximum demand is 85 amps.

The Measured Z_S is 0.48 Ω and an insulation test on all conductors shows >200 M ohms. RCD operating times are 43 ms at 30 mA and 19 ms at 150 mA.

A visual inspection shows no defects and the test instruments are: Low resistance ohm and insulation resistance meter Serial no. 08H46. Earth loop impedance tester Serial no. 076H90. RCD tester Serial number 740026.

1. Complete a schedule of inspections.
2. Complete a schedule of test results.
3. Complete the appropriate certificate.
4. Using the rule of thumb, show if the measured value of Z_S is acceptable. Give a reason/reasons why this measured value is lower than $Z_e + (R_1 + R_2)$.
5. If cooker unit had been a replacement for a unit without a socket outlet, complete the Minor Works Certificate.
6. Describe in detail how an earth loop impedance test would be carried out on this circuit.

Exercise 2

You have installed a new ring circuit consisting of eight socket outlets on the ground floor of a detached house for Mr. Hawth who resides in Crawley, Hampshire.

There is no documentation available for the existing installation, but it appears to be in good condition. There is space available in the existing consumer's unit for a new circuit.

The supply is a TNCS 230 V 50 Hz single phase with an 80 amp supply fuse to BS 1361. Measured values of PSCC and Z_e are 1350A and 0.28 Ω, respectively. On completion of the additional work the maximum demand is 78 amps.

Meter tails are 16 mm^2 copper, the earthing conductor is 10 mm^2 copper, and the main equipotential bonding conductors to the oil and water supplies are 10 mm^2 (correctly connected).

The new circuit is wired in 63 metres of 2.5 mm^2/1.5 mm^2 thermoplastic (pvc) twin and earth cable. Protection is by a 30A BS 3036 semi-enclosed rewirable fuse which has a maximum Z_S value of 1.14 Ω. The consumer's unit, situated in a cupboard in the hall, has a 100A BS EN 61008-1 RCD as a main switch with an IΔn rating of 30 mA. It has a tripping time of 56 ms at its rating and 24 ms at five times its rating. The insulation resistance value is >200 MΩ.

A visual inspection of the new circuit shows no defects, and the test instrument used is a Megger multi-function instrument, serial number CJK 1047.

1. Complete the correct paperwork.
   A few weeks after the circuit has been installed, Mr. Hawth requests that you install an additional twin socket outlet in the Kitchen; this is to be spurred from the new ring circuit. The socket is 10 metres away from the nearest existing outlet.
2. Complete the correct paperwork.
   The spur is to be $2.5\,mm^2/1.5\,mm^2$ twin and earth thermoplastic cable (pvc).

   The resistance of $2.5\,mm^2$ copper is $7.41\,m\Omega$ per metre and the resistance of $1.5\,mm^2$ copper is $12.10\,m\Omega$ per metre.

Exercise 3

Mr. P. Knut, who lives in Nutley, Kent, is moving to a new house in Dover, Kent.

This is an existing building which is 18 years old and now requires the correct inspection and test to be carried out and relevant documents to be completed. Existing documentation is available. The last inspection and test was carried out 6 years earlier along with some additions to the installation. The circuits are as follows.

Circuit	Fuse	Live	CPC	Length	Max Z_S	Points
1. Cooker	30 amp BS 3036	$6\,mm^2$	$2.5\,mm^2$	23 m	1.14	1
2. Ring 1	30 amp BS 3036	2.5	1.5	48	1.14	10
3. Ring 2	30 amp BS 3036	2.5	1.5	53	1.14	8
4. I/H	15 amp BS 3036	1.5	1.00	12	5.58	1
5. Lighting	5 amp BS 3036	1.00	1.00	38	18.5	8
6. Lighting	5 amp BS 3036	1.00	1.00	57	18.5	9

$R_1 + R_2$ per metre @ 20 cents

$6\,mm^2$	3.08 m/ohm metre
$2.5\,mm^2$	7.41 m/ohm metre
$1.5\,mm^2$	12.10 m/ohm metre
$1.0\,mm^2$	18.10 m/ohm metre

The supply is a 230 V 50 Hz TN-S system with a Z_e of 0.58 ohms and a PFC of 900A.

The main fuse is an 80A BS1361. The consumers unit is under the stairs and the main switch is 100A 240 V to BS 5773.

Meter tails are 16 mm^2

Earthing conductor is 6 mm^2.

Equipotential bonding to gas and water is 10 mm^2 and is correctly connected. Plumbing is all in copper tubing and there is no evidence of supplementary bonding.

All circuits have an installation value of >200 M ohms.

Circuit 4 has a dimmer to control the lounge lighting, and there is a shaver socket in the bathroom.

The test instrument used is a Megger multi-function instrument, serial number CJK 1047.

1. Using rule of thumb show if the measured value of Z_S is acceptable.
2. Complete a Periodic Inspection Report and Test Result Schedule for this installation.

Exercise 4

Complete a Schedule of Test Results, with the circuit protective devices selected and fitted in the correct order for good working practice. Omit sections where no details are given.

Circuit description	Phase conductor	CPC Conductor	Circuit length
1. Lighting	1.5 mm^2	1 mm^2	32 metres
2. Ring	2.5 mm^2	1.5 mm^2	68 metres
3. Ring	4 mm^2	1.5 mm^2	72 metres
4. Shower	6 mm^2	2.5 mm^2	14 metres
5. Immersion	2.5 mm^2	1.5 mm^2	17 metres
6. Lighting	1 mm^2	1 mm^2	43 metres

Supply system is TN-S 230 volt measured Z_S is 0.43 Ω, and PFC is 1.2 kA.

All circuits are protected by BS EN 60898 type B circuit breaker.

Exercise 5

Use BS 7671 to find the regulation number indicating where RCDs should be used and what trip rating they should have for:

- Fire protection on farms
- Protection where flammable materials are stored
- Fixed equipment in Zones 1–3 in bathrooms
- Restrictive conductive locations
- TT systems
- Circuits with a high earth fault loop impedance (Z_S)
- Caravan parks
- Sockets likely to supply portable equipment used out doors
- Swimming pools

Questions

1. An insulation resistance test has been carried out on a 6-way consumer's unit. The circuits recorded values of $5.6\,m\Omega$, $8.7\,m\Omega$, $>200\,m\Omega$, $>200\,m\Omega$, $12\,m\Omega$ and $7\,m\Omega$.
   Calculate the total resistance of the installation and state giving reasons whether or not the installation is acceptable.

2. A ring circuit is 54 metres long and is wired in $2.5\,mm^2/1.5\,mm^2$ thermoplastic cable. The protective device is a BS EN 60898 type C device and the Z_e for the installation is 0.24Ω. The resistance of $2.5\,mm^2$ copper is $7.41\,m\Omega$ per metre and $1.5\,mm^2$ copper is $12.1\,m\Omega$ per metre.
   (i) Calculate the Z_S for the circuit.
   (ii) Will the protective device be suitable?

3. An A2 radial circuit is wired in $4\,mm^2$ thermoplastic twin and earth cable. It is 23 metres long. The circuit has on it four twin 13 amp socket outlets. Protection is by a 30A BS 3036 semi-enclosed fuse. Z_e for the installation is $0.6\,\Omega$.
   Socket 1 is 12 metres from the consumer unit, socket 2 is 6 metres from socket 1, and socket 3 is 2.5 metres from socket 2.
   (i) Calculate the $R_1 + R_2$ value at each socket outlet.
   (ii) Will the circuit protective device be suitable?

4. A 9.5 kW electric shower has been installed using $10\,mm^2/4\,mm^2$ thermoplastic twin and earth cable which is 14.75 metres long. The circuit is to be connected to a spare way in the existing consumer's unit.

Protection is by a BS 3036 semi-enclosed 45 amps rewirable fuse. Z_e for the system is $0.7\,\Omega$. The temperature at the time of testing is $20°C$.

(i) Calculate $R_1 + R_2$ for this circuit.

(ii) Will this circuit meet the required disconnection time?

5. A ring circuit is wired $4\,mm^2$ singles in conduit. The circuit is 87 metres in length and is protected by a 32A BS 3871 type 2 circuit breaker, the maximum Z_S permissible is $1.07\,\Omega$ and the actual Z_e is $0.63\,\Omega$. Calculate:

(i) The expected Z_S.

(ii) The maximum permissible length that could be allowed for a spur in $4\,mm^2$ cable.

6. List the certification that would be required after the installation of a new lighting circuit.

7. List three non-statutory documents relating to electrical installation testing.

8. List four reasons why a Periodic Test Report would be required.

9. Apart from a new installation, under which circumstances would a Periodic Inspection Report **not** be required?

10. A ring circuit is wired in $2.5\,mm^2/1.5\,mm^2$. The resistance of the phase and neutral loops were each measured at $0.3\,\Omega$. Calculate the:

(i) The resistance between P and N at each socket after all interconnections have been made.

(ii) End to end resistance of the CPC.

(iii) Resistance between P and CPC at each socket after all interconnections have been made.

11. A spur has been added to the ring circuit in Question 10. The additional length of cable used is 5.8 metres.
Calculate the $R_1 + R_2$ for this circuit.

12. What is a 'statutory' document?

13. What is a 'non-statutory document'?

14. Why is it important to carry out testing on a new installation in the correct sequence?

15. How many special locations are listed in the BS 7671 amended to 2004?

16. State the affect that increasing the length of a conductor could have on its insulation resistance.

17. An installation has seven circuits. Circuits 1, 4 and 6 have insulation resistances of greater than 200 mΩ. Circuits 2, 3, 5 and 7 have resistance values of 50, 80, 60, and 50, respectively. Calculate the total resistance of the circuit.

18. State the correct sequence of tests for a new domestic installation connected to a TT supply.

19. List, in the correct sequence, the instruments required to carry out the tests in Question 18.

20. State the values of the test currents required when testing a 30 mA RCD used for supplementary protection against direct contact.

21. How many times its rated operating current is required to operate a type B BS EN 60898 circuit breaker instantaneously?

22. What is the maximum resistance permitted for equipotential bonding?

23. What would be the resistance of 22 metres of a single 10 mm^2 copper conductor?

24. Which type of supply system uses the mass of earth for its earth fault return path?

25. The Table below shows the resistance values recorded at each socket on a ring circuit during a ring circuit test after the interconnections had been made. Are the values as expected? If not, what could the problem be? The temperature is 20°C; the end to end resistances of the conductors are: Phase 0.45 Ω, neutral 0.46 Ω and CPC is 0.75 Ω.

Socket	P to N	P to CPC
1	0.225	0.35
2	No reading	No reading
3	0.224	No reading
4	No reading	0.35
5	0.34	0.50
6	0.4	0.35
7	0.22	0.35

26. A lighting circuit is to be wired in $1\,mm^2$ twin and earth thermoplastic cable, the circuit is protected by a 5A BS 3036 fuse. What would be the maximum length of cable permissible to comply with the earth fault loop impedance requirements (Z_S)? (Z_e is 0.45.)

27. With regard to the *On-Site Guide* what are the stated earth loop impedance values outside of a consumer's installation for a TT, TNS, and TNCS supply?

28. A ring final circuit has twelve twin 13 amp socket outlets on it. How many unfused spurs would it be permissible to add to this circuit?

29. How many fused spurs would it be permissible to connect to the ring circuit in Question 28?

30. Name the document that details the requirements for electrical test equipment.

31. State three extraneous conductive parts that could be found within a domestic installation.

32. State four exposed conductive parts commonly found within an electrical installation.

33. State the minimum c.s.a. for a non-mechanically protected, supplementary bonding conductor that could be used in a bathroom.

34. What is the minimum acceptable insulation resistance value permissible for a complete 400 V a.c. 50 Hz installation?

35. State the test voltage and current required for an insulation test carried out on a 230 V a.c. 50 Hz installation.

36. The Electricity at Work Regulations state that for a person to be competent when carrying out inspecting and testing on an electrical installation they must be .... What?
   (i)   In possession of technical knowledge.
   (ii)   Experienced, or
   (iii) Supervised.

37. A 400 V a.c. installation must be tested with an insulation resistance tester set at .... volts.

38. List three requirements of GS 38 for test leads.

39. List three requirements of GS 38 for test probes.

40. Name a suitable piece of equipment that could be used for testing for the presence of voltage while carrying out the isolation procedure.

41. To comply with BS 7671 the purpose of the initial verification is to verify that...............

42. To comply with GN 3, what are the four responsibilities of the inspector?

43. When testing a new installation, a fault is detected on a circuit. State the procedure that should be carried out.

44. State three reasons for carrying out a polarity test on a single phase installation.

45. On which type of ES lampholders is it **not** necessary to carry out a polarity test?

46. What is the minimum requirement of BS 7671 for ingress protection of electrical enclosures?

47. A 6A Type B circuit breaker trips each time an earth loop impedance test is carried out on its circuit. How could the Z_S value for this circuit be obtained?

48. What is the maximum rating permissible for before a motor would require overload protection?

49. Identify the type of circuit breaker that should be used for:
    (a) Discharge lighting in a factory.
    (b) A large transformer.
    (c) A 3 phase motor.

50. Identify three warning labels and notices that could be found in an installation.

51. The circuits in the table have been tested and the earth fault loop impedance values for each circuit are as shown. Using the rule-of-thumb method to identify whether the circuits will comply with BS 7671.

Measured Z_S	Maximum Z_S
0.86 Ω	1.2 Ω
0.68 Ω	0.96 Ω
1.18 Ω	1.5 Ω
2.8 Ω	4 Ω
1.75 Ω	2.4 Ω

52. State the minimum IP rating for fixed equipment in Zone 2 of a bathroom.

53. State the minimum size for an equipotential bonding conductor installed in a TNS system with $25\,\text{mm}^2$ metre tails.

54. Identify the documentation that should be completed after the installation of a cooker circuit.

55. State four non-statutory documents.

56. State four statutory documents.

57. State the sequence of colours for a new 3 phase and neutral system.

Answers to Exercises

(*The Certificates can be found at the end of the chapter*)

Exercise 1

1. and 2. Part P – Domestic Electrical Installation Certificate – Part 2.
3. Part P – Domestic Electrical Installation Certificate.
4. $1.92 \times 0.75 = 1.44\ \Omega$. This is the maximum permissible Z_S for a BS 88 32 amp fuse. As the measured value of Z_S was $0.48\ \Omega$ this circuit is fine.
    The measured Z_S is lower due to parallel paths through the bonding conductors, etc.
5. Minor Domestic Electrical Installation Works Certificate.
6. Use an earth fault loop impedance test instrument set on loop, plug lead into socket outlet on cooker control unit and take earth loop impedance reading.

Exercise 2

1. Part P – Domestic Electrical Installation Certificate; Part P – Domestic Electrical Installation Certificate – Part 2.
2. Minor Domestic Electrical Installation Works Certificate.

Exercise 3

1. Rule-of-thumb

Circ no.	Max Z_S	Calculated	Actual	✗/✓
1	$1.14 \times 0.75 =$	0.85	0.82	ok
2	$1.14 \times 0.75 =$	0.85	0.81	ok
3	$1.14 \times 0.75 =$	0.85	0.83	ok
4	$5.58 \times 0.75 =$	4.18	0.94	ok
5	$18.5 \times 0.75 =$	13.87	1.95	ok
6	$18.5 \times 0.75 =$	13.87	2.64	ok

2. Periodic inspection and test reports. Periodic Inspection Report for an Electrical Installation; Supply Characteristics and Earthing Arrangements; Survey and Test Report Schedule.
   (a) There is no supplementary bonding in the bathroom.
   (b) The earthing conductor is undersized.
   (c) RCD protection is advisable for the ground floor ring circuit although it is not a requirement as the wiring regulations are not retrospective.

The inspection checklist indicates '✗' for presence of RCD for supplementary protection as there is a downstairs ring circuit which should have one to comply with current regulations, this also requires a '✗' for particular protective measures for special locations.

Electrical separation has a tick as there is a shaver socket in the bathroom.

A tick ✓ is required in segregation of band I and II circuits where the installation has circuits such as bell, telephone and TV aerials, etc.

Exercise 4

Part P – Domestic Electrical Installation Certificate – Part 2.

Exercise 5

(a) Used for fire protection 500 mA (Regulation 605-10-01).

(b) Particular risk of danger of fire exist 300 mA (Regulation 482-02-06).

(c) Bathroom is a special location, Regulation 601-09 requires the use of a 30 mA RCD.

(d) An RCD is not permitted as a form of protection in a restrictive conductive location.

(e) TT system should preferably use a split board with RCD rated at 100 mA for fixed equipment and 30 mA for socket outlets (*On-Site Guide*).

(f) Where a circuit has high loop impedance a 100 mA RCD is suitable providing it is not for socket outlets that could be used outside and that the circuit is not in a special location.

(g) Caravan park is a special location, Regulation 608-13-05 requires the use of a 30 mA RCD.

(h) Regulation 471-16-01 refers to Regulation 412-06-02 which requires the use of a 30 mA RCD.

(i) Regulation 602-04-01 (ii) refers to Regulation 412-06-02 which requires the use of a 30 mA RCD installed in accordance with Regulations 602-07-01 and 602-07-02 (iii).

DOMESTIC ELECTRICAL INSTALLATION CERTIFICATE

Issued in accordance with *British Standard 7671 – Requirements for Electrical Installations* by a Domestic Installer registered with NICEIC, Warwick House, Houghton Hall Park, Houghton Regis, Dunstable, LU5 5ZX

This certificate is not valid if the serial number has **DCP3/** been defaced or altered

TRACEABLE SERIAL NUMBER

nic DOMESTIC INSTALLER

This safety certificate is an important and valuable document which should be retained for future reference

DETAILS OF THE CLIENT

Client and address

MRS F.G. GRANT
3 BISHOPS CLOSE
BATH
SOMERSET   Postcode SO3 6HT

ADDRESS OF THE INSTALLATION

Installation address

3 BISHOPS CLOSE
BATH
SOMERSET   Postcode SO3 6HT

The installation is

New
An addition
An alteration

DETAILS OF THE INSTALLATION

Extent of the installation work covered by this certificate

COOKER CIRCUIT

DESIGN, CONSTRUCTION, INSPECTION AND TESTING

I/We being the person(s) responsible for the design, construction, inspection and testing of the electrical installation (as indicated by my/our signatures adjacent), particulars of which are described above, having exercised reasonable skill and care when carrying out the design, construction, inspection and testing, hereby CERTIFY that the said work for which I/we have been responsible is to the best of my/our knowledge and belief, in accordance with BS 7671, 200I amended to 200+ (date) except for the departures, if any, detailed as follows:

Details of departures from BS 7671, as amended (Regulations 120-01-03, 120-02):

NONE

The extent of liability of the signatory is limited to the work described above as the subject of this certificate. For the **DESIGN**, the **CONSTRUCTION** and the **INSPECTION AND TESTING** of the installation.

Signature R ROGERS   Name (CAPITALS) R ROGERS   Date 1-07-08

The results of the inspection and testing reviewed by the **Qualified Supervisor**

Signature T Rigger   Name (CAPITALS) T. RIGGER   Date 2-07-08

NEXT INSPECTION

I RECOMMEND that this installation is further inspected and tested after an interval of not more than $ 10 YRS

§ Enter interval in terms of years, months or weeks, as appropriate

COMMENTS ON EXISTING INSTALLATION

NONE

Note: Enter 'NONE' or, where appropriate, the page number(s) of additional page(s) of comments on the existing installation

PARTICULARS OF THE DOMESTIC INSTALLER

Trading title   R AND T ELECTRICAL

Address   6 COWBOY CLOSE
BATH
SOMERSET   Postcode SO3 6QT

Telephone No 01271 246810

NICEIC Registration No (Essential information)   1 0 9 8 7 6 5 4 3

SCHEDULE OF ADDITIONAL RECORDS*

NONE

See attached schedule

Please see the 'Notes for Recipients' on the reverse of this page.

Page 1 of 3

* Where the electrical work to which this certificate relates includes the installation of a fire alarm system and/or an emergency lighting system (or a part of such system), this electrical safety certificate should be accompanied by the particular certificate(s) for the system(s)

DCPyl

Original (To the person ordering the work)

DOMESTIC ELECTRICAL INSTALLATION CERTIFICATE

DOMESTIC INSTALLER

SUPPLY CHARACTERISTICS
Tick boxes and enter details, as appropriate

System type(s)

TN-S	✓
TN-C-S	
TT	

Number and type of live conductors

1-phase (2 wire)	✓
1-phase (3 wire)	
3-phase (3 wire)	
3-phase (4 wire)	
Other	

Phase state

Nature of supply parameters
Notes: (1) by enquiry (2) by enquiry or by measurement (3) by measurement. (3) where more than one supply, record the higher or highest values

Single-phase

Nominal voltage(s) $U^{(1)}$: ✗ V

Nominal frequency, $f^{(1)}$: 50 Hz

$U_o^{(1)}$: 230 V

Prospective fault current, $I_{pf}^{(2)(3)}$: 0·8 kA ✗

3-phase

Prospective fault current, $I_{pf}^{(2)(3)}$: N/A kA ✗

External earth fault loop impedance, Z_e, $f^{(1)}$: 0·24 Ω

Characteristics of primary supply overcurrent protective device(s)

BS(EN): 1361

Type: 1

Nominal current rating: 100 A

Short-circuit capacity: 16·5 kA

Main switch or circuit-breaker

Type BS(EN): 60439

No of poles: 2

Supply conductors material: Cu

Supply conductors csa: 25 mm²

Voltage rating: 230 V

Current rating, I_n: 100 A

RCD operating current $I_{\Delta n}$: N/A mA

RCD operating time (at $I_{\Delta n}$)*: N/A ms

applicable only where an RCD is used as a main circuit-breaker

PARTICULARS OF INSTALLATION AT THE ORIGIN
Tick boxes and enter details, as appropriate

Means of earthing

Distributor's facility	✓
Installation earth electrode	N/A

Details of installation earth electrode (where applicable)

Type (eg rod(s), tape etc): N/A

Location: N/A

Electrode resistance, R_A: N/A Ω

Method of measurement: N/A

Earthing conductor

Conductor material: [blank]

Conductor csa: [blank] mm²

Continuity check: ✓

Measured Z_e: 0·24 Ω

Maximum demand (Load): 85 A per phase

Number of smoke alarms: NONE

Main equipotential bonding conductors and bonding of extraneous-conductive-parts (✓)

	Conductor material	Conductor csa	Location (where not obvious)
Water service		[blank] mm²	
Gas service			
Oil service			
Structural steel			
Other incoming service(s)			

SCHEDULE OF ITEMS INSPECTED
† *See note below*

Methods of protection against electric shock
- ✓ Insulation of live parts, and barriers or enclosures
- N/A Presence of RCD(s) for supplementary protection against direct contact and/or protection against indirect contact
- ✓ Presence of earthing conductor and circuit protective conductors
- ✓ Presence of main equipotential bonding conductors
- N/A Presence of supplementary equipotential bonding conductors
- N/A Class II fixed equipment
- N/A SELV

Methods of protection against electric shock *(cont)*
- N/A Electrical separation

Prevention of mutual detrimental influence
- ✓ Proximity of non-electrical services and other influences
- ✓ Segregation of Band I and Band II circuits or Band II insulation used

Identification
- ✓ Presence of diagrams, instructions, circuit charts and similar information
- ✓ Presence of danger notices
- ✓ Presence of other warning notices, including presence of mixed wiring colours

Identification *(cont)*
- ✓ Labelling of protective devices, switches and terminals
- ✓ Identification of conductors

Cables and conductors
- ✓ Routing of cables in prescribed zones or within mechanical protection
- ✓ Connection of conductors
- ✓ Erection methods
- ✓ Selection of conductors for current-carrying capacity and voltage drop
- ✓ Presence of fire barriers, suitable seals and protection against thermal effects

General
- ✓ Presence and correct location of appropriate devices for isolation and switching
- ✓ Adequacy of access to switchgear and other equipment
- N/A Particular protective measures for special installations and locations
- ✓ Connection of single-pole devices for protection or switching in phase conductors only
- ✓ Correct connection of accessories and equipment
- ✓ Choice and setting of protective devices (for protection against indirect contact and/or overcurrent)
- ✓ Selection of equipment and protective measures appropriate to external influences
- ✓ Selection of appropriate functional switching devices

SCHEDULE OF ITEMS TESTED
† *See note below*

- ✓ External earth fault loop impedance, Z_e
- N/A Installation earth electrode resistance, R_A
- ✓ Continuity of protective conductors
- N/A Continuity of ring final circuit conductors
- ✓ Insulation resistance between live conductors
- ✓ Insulation resistance between live conductors and earth
- ✓ Polarity
- ✓ Earth fault loop impedance, Z_s
- ✓ Operation of residual current device(s)
- ✓ Functional testing of assemblies

† *All boxes must be completed.* '✓' indicates that an inspection or a test was carried out and that the result was **satisfactory**. 'N/A' indicates that an inspection or test was **not applicable** to the particular installation.

This form is based on the model Electrical Installation Certificate shown in Appendix 6 of BS 7671 (as amended).
Published by NICEIC Group Limited © Copyright Electrical Safety Council (July 2006)

Page 2 of 3

DCP3 / 3

DOMESTIC ELECTRICAL INSTALLATION CERTIFICATE

NICEIC DOMESTIC INSTALLER

CIRCUIT DETAILS / TEST RESULTS

CODES FOR TYPE OF WIRING

Code	Wiring type
A	PVC/PVC cables
B	PVC cables in metallic conduit
C	PVC cables in non-metallic conduit
D	PVC cables in metallic trunking
E	PVC cables in non-metallic trunking
F	PVC/SWA cables
G	XLPE/SWA cables
H	Mineral-insulated cables
O	Other - please state

Circuit number	Circuit designation	Type of circuit (D/F)	Reference method	Number of points served	Live (mm²)	cpc (mm²)	Max. disconnection time permitted by BS 7671	BS (EN)	Type No	Rating (A)	Capacity (kA)	RCD Operating current (mA)	Maximum Zs permitted by BS 7671 (Ω)	r_1 Phase	r_n Neutral	r_2 cpc	R_1+R_2	R_2	Phase/Phase	Phase/Neutral	Phase/Earth	Neutral/Earth	Polarity ✓	Maximum measured earth fault loop impedance Z_s (Ω)	at $I_{\Delta n}$ ms	at $5 I_{\Delta n}$ ms
*																										
1	COOKER	F	A	6	1	6		25 04	6009	B	32	6	30	1.5				0.35		>200	>200	>200		✓	0.48	43	19

Location of consumer units)

Designation of consumer unit(s)

Prospective fault current at consumer unit(s) ____ kA

TEST INSTRUMENTS

Test instruments (serial numbers) used

Multi-functional	Continuity	Insulation resistance	Earth fault loop impedance	RCD	Earth electrode resistance
N/A	08H46	08H46	076H90	740026	N/A

DCP1/5

NICEIC
DOMESTIC INSTALLER

This safety certificate is an important and valuable document which should be retained for future reference

This certificate is not valid if the serial number has been defaced or altered

DMP2/

TRACEABLE SERIAL NUMBER

MINOR DOMESTIC ELECTRICAL INSTALLATION WORKS CERTIFICATE

Issued in accordance with *British Standard 7671 – Requirements for Electrical Installations* by an Approved Contractor or Conforming Body enrolled with NICEIC, Warwick House, Houghton Hall Park, Houghton Regis, Dunstable, LU5 5ZX.

To be used only for minor electrical work which does not include the provision of a new circuit

PART 1: DETAILS OF THE MINOR WORKS

Client: **MRS F G GRANT**

Date minor works completed: **01-07-08**

Contract reference, if any: —

Description of the minor works: **REPLACE EXISTING COOKER UNIT FOR UNIT WITH 13AMP SOCKET OUTLET**

Details of departures, if any, from BS 7671 (as amended): **NONE**

Location/address of the minor works: **3 BISHOPS CLOSE BATH SOMERSET** Postcode **SO3 6HT**

PART 2: DETAILS OF THE MODIFIED CIRCUIT

System type and earthing arrangements: TN-C-S ☐ TN-S ✓ TT ☐

Method of protection against indirect contact: EEBAD ✓ Other ☐

Overcurrent protective device for the modified circuit: BS(EN) **61009** Type **B** Rating **32** A

Residual current device (if applicable): BS(EN) **61009** Type **B** $I_{\Delta n}$ **30** mA

Details of wiring system used to modify the circuit Type **PVC/PVC** Reference method **6** csa of live conductors **6** mm² csa of cpc **2·5** mm²

Where protection against indirect contact is EEBAD, Maximum disconnection time permitted by BS 7671 **0·4** s Maximum Z_s permitted by BS 7671 **1·5** Ω

Comments, if any, on existing installation: **NONE**

PART 3: INSPECTION AND TESTING OF THE MODIFIED CIRCUIT AND RELATED PARTS †*Essential inspections and tests*

† Confirmation that necessary inspections have been undertaken ✓ (✓)

† Circuit resistance $R_1 + R_2$ **0·33** Ω or R_2 — Ω

Insulation resistance (* In a multi-phase circuit, record the lower or lowest value, as appropriate)

Phase/Phase* — MΩ

Phase/Neutral* **>200** MΩ

Instrument Serial No(s). **08H46 / 076H90 / 740026**

† Phase/Earth* **>200** MΩ

† Neutral/Earth **>200** MΩ

† Confirmation of the adequacy of earthing ✓ (✓)

† Confirmation of the adequacy of equipotential bonding ✓ (✓)

† Confirmation of correct polarity ✓ (✓)

† Maximum measured earth fault loop impedance, Z_s **0·48** Ω

† RCD operating time at $I_{\Delta n}$ (if RCD fitted) **43** ms

RCD operating time at 5 x $I_{\Delta n}$, if applicable **19** ms

Agreed limitations, if any, on the inspection and testing: **COOKER CIRCUIT ONLY**

PART 4: DECLARATION

I/We CERTIFY that the said works do not impair the safety of the existing installation, that the said works have been designed, constructed, inspected and tested in accordance with BS 7671: **2001** (IEE Wiring Regulations), amended to **2004** and that the said works, to the best of my/our knowledge and belief, at the time of my/our inspection complied with BS 7671 except as detailed in Part 1.

The results of the inspection and testing reviewed by the Qualified Supervisor

For and on behalf of *(Trading Title of Domestic Installer)* **R AND T ELECTRICS**

Name (CAPITALS) **R ROGERS**

Name (CAPITALS) **T RIGGER**

Address **6 COWBOY CLOSE BATH SOMERSET** Postcode **SO3 6QT**

Signature **R Rogers**

Signature **T Rigger**

Date **1-07-08**

Date **1-01-08**

Registration Number **1 0 9 8 7 6 5 4 3** *(The registration number is essential information)*

This form is based on the model shown in Appendix 6 of BS 7671 (as amended)
Published by the NICEIC Group Limited © Copyright The Electrical Safety Council (July 2006)

Please see the 'Notes for Recipients' on the reverse of this page.

DMP2/1

This certificate is not valid
if the serial number has **DCP3/**
been defaced or altered

TRACEABLE
SERIAL NUMBER

DOMESTIC
INSTALLER

DOMESTIC ELECTRICAL INSTALLATION CERTIFICATE

Issued in accordance with *British Standard 7671 – Requirements for Electrical Installations* by a Domestic Installer registered with NICEIC.
Warwick House, Houghton Hall Park, Houghton Regis, Dunstable, LU5 5ZX

This safety certificate is an important and
valuable document which should be retained
for future reference

DETAILS OF THE CLIENT

Client and address

MR HAWTH
22 NORMANS AVENUE
CRAWLEY
HAMPSHIRE

Postcode SO21 Y33

ADDRESS OF THE INSTALLATION

Installation address

22 NORMANS AVENUE
CRAWLEY
HAMPSHIRE

Postcode SO21 Y33

The installation is

New

An addition ✓

An alteration

DETAILS OF THE INSTALLATION

Extent of the installation work covered by this certificate

NEW RING CIRCUIT ON GROUND FLOOR

DESIGN, CONSTRUCTION, INSPECTION AND TESTING

I/We being the person(s) responsible for the design, construction, inspection and testing of the electrical installation (as indicated by my/our signatures adjacent), particulars of which are described above, having exercised reasonable skill and care when carrying out the design, construction, inspection and testing, hereby CERTIFY that the said work for which I/we have been responsible is to the best of my/our knowledge and belief, in accordance with BS 7671, 2001 amended to 2004 (date) except for the departures, if any, detailed as follows:

Details of departures from BS 7671, as amended (Regulations 120-01-03, 120-02):

NONE

The extent of liability of the signatory is limited to the work described above as the subject of this certificate.
For the **DESIGN**, the **CONSTRUCTION** and the **INSPECTION AND TESTING** of the installation.

Signature R Rogers   Name (CAPITALS) R ROGERS   Date 01-07-08

The results of the inspection and testing reviewed by the Qualified Supervisor

Signature T Rigger   Name (CAPITALS) T RIGGER   Date 01-07-08

PARTICULARS OF THE DOMESTIC INSTALLER

Trading title   R AND T ELECTRICAL

Address   6 COWBOY CLOSE
BATH
SOMERSET

Postcode SO3 6DE

NICEIC Registration No 1 0 9 8 7 6 5 4 3

Telephone No (Essential information)

NEXT INSPECTION

§ Enter interval in terms of years, months or weeks, as appropriate

I RECOMMEND that this installation is further inspected and tested after an interval of not more than § 10

COMMENTS ON EXISTING INSTALLATION

Note: Enter 'NONE or', where appropriate, the page number(s) of additional page(s) of comments on the existing installation

NO DOCUMENTATION AVAILABLE FOR EXISTING INSTALLATION. APPEARS TO BE IN GOOD CONDITION

SCHEDULE OF ADDITIONAL RECORDS*

See attached schedule

N/A

Please see the "Notes for Recipients" on the reverse of this page.

Page 1 of 3

* Where the electrical work to which this certificate relates includes the installation of a fire alarm system and/or an emergency lighting system (or a part of such systems), this electrical safety certificate should be accompanied by the particular certificate(s) for the system(s) shown in Appendix 6 of BS 7671 (as amended).

DCP3/1

DOMESTIC ELECTRICAL INSTALLATION CERTIFICATE

This certificate is not valid if the serial number has been defaced or altered

DCP3/

TRACEABLE SERIAL NUMBER

SUPPLY CHARACTERISTICS
Tick boxes and enter details, as appropriate

System type(s)

Number and type of live conductors

TN-S	
TN-C-S	✓
TT	

1-phase (2 wire)	✓
3-phase (3 wire)	
Other	

1-phase (3 wire)	
3-phase (4 wire)	

Phase state

Nature of supply parameters
Notes: (1) by enquiry (2) by enquiry or by measurement (3) where more than one supply, record the higher or highest values

Nominal U $^{(1)}$ voltage(s)	—	V
Nominal frequency, f $^{(1)}$	50	Hz
U_o $^{(1)}$	230	V
External earth fault loop impedance, Z_e $^{(2)}$	0.28	Ω

Single-phase

Prospective fault current, I_{pf} $^{(2)(3)}$	1.35	kA

3-phase

Prospective fault current, I_{pf} $^{(2)(3)}$		kA

Characteristics of primary supply overcurrent protective device(s)

BS(EN)	3036	
Type	N/A	
Nominal current rating	30	A
Short-circuit capacity		kA

Main switch or circuit-breaker

Type BS(EN)	60081	
No of poles	2	
Supply conductors material	Cu	
Supply conductors csa	16	mm²
Voltage rating	230	V
Current rating, I_n	100	A
RCD operating current, $I_{\Delta n}$*	56	mA
RCD operating time (at $I_{\Delta n}$)*	24	ms

*applicable only where an RCD is used as a main circuit-breaker

PARTICULARS OF INSTALLATION AT THE ORIGIN
Tick boxes and enter details, as appropriate

Means of earthing

Distributor's facility	✓
Installation earth electrode	N/A

Details of installation earth electrode (where applicable)

Type (eg rod(s), tape etc)	N/A
Electrode resistance, R_A	N/A Ω
Location	N/A
Method of measurement	

Earthing conductor

Conductor material	COPPER
Conductor csa	10 mm²
Continuity check	✓

Conductor material	Cu
Location (where not obvious)	

Main equipotential bonding conductors and bonding of extraneous-conductive-parts (✓)

Conductor csa	10 mm²
Water service	✓
Oil service	✓
Structural steel	N/A
Gas service	N/A
Other incoming service(s)	N/A

Measured Z_s: 0.28 Ω A per phase

Maximum demand (Load): 78

Number of smoke alarms: NONE

SCHEDULE OF ITEMS INSPECTED
† See note below

Methods of protection against electric shock

Insulation of live parts, and barriers or enclosures	✓
Presence of RCD(s) for supplementary protection against direct contact and/or protection against indirect contact	✓
Presence of earthing conductor and circuit protective conductors	✓
Presence of main equipotential bonding conductors	✓
Presence of supplementary equipotential bonding conductors	N/A
Class II fixed equipment	N/A
SELV	N/A

Methods of protection against electric shock
† See note below

Electrical separation	N/A

Prevention of mutual detrimental influence

Proximity of non-Electrical Services and other influences	✓
Segregation of Band I and Band II circuits or Band II insulation used	✓

Identification

Presence of diagrams, instructions, circuit charts and similar information	✓
Presence of danger notices	✓
Presence of other warning notices, including presence of mixed wiring colours	✓

Identification (cont)

Labelling of protective devices, switches and terminals	✓
Identification of conductors	✓

Cables and conductors

Routing of cables in prescribed zones or within mechanical protection	✓
Connection of conductors	✓
Erection methods	✓
Selection of conductors for current-carrying capacity and voltage drop	✓
Presence of fire barriers, suitable seals and protection against thermal effects	✓

General

Presence and correct location of appropriate devices for isolation and switching	✓
Adequacy of access to switchgear and other equipment	✓
Particular protective measures for special installations and locations	N/A
Connection of single-pole devices for protection or switching in phase conductors only	✓
Correct connection of accessories and equipment	✓
Choice and setting of protective devices (for protection against indirect contact and/or overcurrent)	✓
Selection of equipment and protective measures appropriate to external influences	✓
Selection of appropriate functional switching devices	✓

SCHEDULE OF ITEMS TESTED
† See note below

External earth fault loop impedance, Z_s	✓
Installation earth electrode resistance, R_A	N/A
Continuity of protective conductors	✓
Continuity of ring final circuit conductors	✓
Insulation resistance between live conductors	✓
Insulation resistance between live conductors and earth	✓
Polarity	✓
Earth fault loop impedance, Z_s	✓
Operation of residual current device(s)	✓
Functional testing of assemblies	✓

† All boxes must be completed. '✓' indicates that an inspection or a test was carried out and that the result was **satisfactory**. 'N/A' indicates that an inspection or test was **not applicable** to the particular installation.

This form is based on the model Electrical Installation Certificate shown in Appendix 6 of BS 7671 (as amended).
Published by NICEIC Group Limited © Copyright Electrical Safety Council (July 2006)

NICEIC DOMESTIC INSTALLER

DOMESTIC ELECTRICAL INSTALLATION CERTIFICATE

DOMESTIC INSTALLER

CIRCUIT DETAILS

TEST RESULTS

CODES FOR TYPE OF WIRING	
A	PVC/PVC cables
B	PVC cables in metallic conduit
C	PVC cables in non-metallic conduit
D	PVC cables in metallic trunking
E	PVC cables in non-metallic trunking
F	XLPE/SWA cables
G	PVC/SWA cables
H	Mineral-insulated cables
O	O (Other - please state)

Circuit number	Circuit designation	Type of wiring (see code)	Reference method of BS 7671 (see Appendix 4)	Number of points served	Live mm²	cpc mm²	Max. disconnection time permitted by BS 7671 s	BS (EN)	Type No	Rating A	Capacity kA	Operating current IΔn mA	Maximum Zs permitted by BS 7671 Ω	t₁ Phase	t₂ Neutral	t₂ cpc	R₁+R₂	R₂	Phase/Phase MΩ	Phase/Neutral MΩ	Phase/Earth MΩ	Neutral/Earth MΩ	Polarity ✓	Maximum measured earth fault loop impedance Zₛ Ω	at IΔn ms	at 5 IΔn (if applicable) ms
1	GROUND FLOOR RING	F	A	6	8	2.5	1.5	50.4	3036		30			301	1.40	0.40	0.46	0.80	0.08	>1000	>1000	>1000	✓	0.3756	24	

Overcurrent protective devices • Circuit conductors csa • Insulation resistance • RCD operating times

* To be completed only where this consumer unit is remote from the origin of the installation. Record details of the circuit supplying this consumer unit in the bold box.
* D = Distribution circuit, F = Final circuit

Location of consumer unit(s) HALL CUPBOARD

Designation of consumer unit(s) 1

Prospective fault current at consumer unit(s) 1·35 kA

TEST INSTRUMENTS

Test instruments (serial numbers) used

Multi-functional CJK1047

Insulation resistance

Continuity

Earth electrode resistance

Earth fault loop impedance

RCD

This certificate is not valid if the serial number has been defaced or altered

DMP2/

TRACEABLE SERIAL NUMBER

MINOR DOMESTIC ELECTRICAL INSTALLATION WORKS CERTIFICATE

Issued in accordance with *British Standard 7671 – Requirements for Electrical Installations* by an Approved Contractor or Conforming Body enrolled with NICEIC, Warwick House, Houghton Hall Park, Houghton Regis, Dunstable, LU5 5ZX.

Original (To the person ordering the work)

To be used only for minor electrical work which does not include the provision of a new circuit

PART 1: DETAILS OF THE MINOR WORKS

Client: MR HAWTH

Date minor works completed: 3-8-08   Contract reference, if any: N/A

Description of the minor works: ADDITIONAL TWIN SOCKET OUTLET IN KITCHEN

Details of departures, if any, from BS 7671 (as amended): NONE

Location/address of the minor works: 22 NORMANS AVENUE CRAWLEY HAMPSHIRE   Postcode: SO21 Y33

PART 2: DETAILS OF THE MODIFIED CIRCUIT

System type and earthing arrangements: TN-C-S ✓   TN-S ☐   TT ☐

Method of protection against indirect contact: EEBAD ☐   Other ☐

Overcurrent protective device for the modified circuit: BS(EN) 2036   Type N/A   Rating 30 A

Residual current device (if applicable): BS(EN) 61008-1   Type N/A $I_{\Delta n}$ 30 mA

Details of wiring system used to modify the circuit: Type PVC/PVC   Reference method 6   csa of live conductors 2.5 mm²   csa of cpc 1.5 mm²

Where protection against indirect contact is EEBAD Maximum disconnection time permitted by BS 7671: 0.4 s   Maximum Z_s permitted by BS 7671: 1.14 Ω

Comments, if any, on existing installation: INSTALLATION APPEARS TO BE IN GOOD CONDITION. WOULD ADVISE THAT A SURVEY SHOULD BE CARRIED OUT TO PREPARE DOCUMENTATION

PART 3: INSPECTION AND TESTING OF THE MODIFIED CIRCUIT AND RELATED PARTS † *Essential inspections and tests*

† Confirmation that necessary inspections have been undertaken ✓ (✓)

† Circuit resistance $R_1 + R_2$ 0.28 Ω   or R_2 N/A Ω

Insulation resistance (* In a multi-phase circuit, record the lower or lowest value, as appropriate)   Phase/Phase* N/A MΩ

Instrument Serial No(s). CJK 1047   † Phase/Neutral* >200 MΩ   † Phase/Earth* >200 MΩ   † Neutral/Earth >200 MΩ

† Confirmation of the adequacy of earthing ✓ (✓)

† Confirmation of the adequacy of equipotential bonding ✓ (✓)

† Confirmation of correct polarity ✓ (✓)

† Maximum measured earth fault loop impedance, Z_s 0.56 Ω

† RCD operating time at $I_{\Delta n}$ (if RCD fitted) 24 ms

RCD operating time at 5 x $I_{\Delta n}$, if applicable 56 ms

Agreed limitations, if any, on the inspection and testing: RING CIRCUIT ONLY

PART 4: DECLARATION

I/We CERTIFY that the said works do not impair the safety of the existing installation, that the said works have been designed, constructed, inspected and tested in accordance with BS 7671: 2001 (IEE Wiring Regulations), amended to 2004 and that the said works, to the best of my/our knowledge and belief, at the time of my/our inspection complied with BS 7671 except as detailed in Part 1.

Name (CAPITALS): R ROGERS

Signature: R Rogers

Date: 3-8-08

The results of the inspection and testing reviewed by the Qualified Supervisor

Name (CAPITALS): T RIGGER

Signature: T Rigger

Date: 4-8-08

For and on behalf of (Trading Title of Domestic Installer): R AND T ELECTRICAL

Address: 6 COWBOY CLOSE BATH SOMERSET   Postcode: SO36QF

Registration Number: 1 0 9 8 7 6 5 4 3   (The registration number is essential information)

This form is based on the model shown in Appendix 6 of BS 7671 (as amended)
Published by the NICEIC Group Limited © Copyright The Electrical Safety Council (July 2006)

Please see the 'Notes for Recipients' on the reverse of this page.

DMP2/1

This report is not valid if the serial number has been defaced or altered

DPP1/

TRACEABLE SERIAL NUMBER

DOMESTIC ELECTRICAL INSTALLATION PERIODIC INSPECTION REPORT (FOR A SINGLE DWELLING)

Issued in accordance with *British Standard 7671–Requirements for Electrical Installations* by a Domestic Installer registered with NICEIC, Warwick House, Houghton Hall Park, Houghton Regis Dunstable LU5 5ZX.

NICEIC DOMESTIC INSTALLER

A DETAILS OF THE CLIENT

Client/Address:
MR. P. KNUT
KERNAL HOUSE
BRAZIL CLOSE
NUTLEY
KENT KT3 8J1

B ADDRESS AND DETAILS OF THE INSTALLATION

Address:
24 THE WARREN
DOVER
KENT
KH10 1DR

Estimated age of the electrical installation: 18 years

If yes, estimated age: 6 years

Evidence of alterations or additions: YES

Electrical Installation Certificate number or previous Periodic Inspection Report number: PIR1047

Records held by: MR KNUT

Date of previous inspection: 2001

Records of installation available: YES

C PURPOSE OF THE REPORT † (see note below)

Purpose for which this report is required:
CLIENT REQUEST

D EXTENT OF THE INSTALLATION AND LIMITATIONS OF THE INSPECTION AND TESTING

Extent of the electrical installation covered by this report:
COMPLETE INSTALLATION

Agreed limitations, if any, on the inspection and testing:
ACCESSIBLE PARTS OF ALL CIRCUITS INSPECTED
ALL CIRCUITS TESTED

E PARTICULARS OF THE DOMESTIC INSTALLER

Trading Title: N. O. SPARKS

Address:
10 HAZELNUT ROAD
DOVER
KENT
Postcode KH10 2NR

Registration Number: 4 1 0 5 1 0 9 8 7
(The registration number is essential information)

F DECLARATION

I/We, being the person(s) responsible for the inspection and testing of the electrical installation (as indicated by my/our signatures below), particulars of which are described above (see B), having exercised reasonable skill and care when carrying out the inspection and testing, hereby declare that the information in this report, including the observations (see G) and the attached schedules (see K and L), provides an accurate assessment of the condition of the electrical installation taking into account the stated extent of the installation and the limitations of the inspection and testing (see D).

I/We further declare that in my/our judgement, the said installation was overall in * _____ condition (see H) at the time the inspection was carried out, and that it should be further inspected as recommended (see I).

* Insert 'a satisfactory' or 'an unsatisfactory', as appropriate

INSPECTION, TESTING AND ASSESSMENT BY:

Signature: N Utley
Name (CAPITALS): N. UTLEY
Position: ELECTRICIAN
Date: 30-6-07

REPORT REVIEWED AND CONFIRMED BY: * See note below

Signature: O Dear
Name (CAPITALS): O. DEAR
Registered Qualified Supervisor for the Domestic Installer at E)
Date: 31-6-07

Please see the 'Notes for Recipients' on the reverse of this page.

Page 1 of 3

† This Domestic Periodic Inspection Report must be used only for reporting on the condition of an existing installation.

‡ The inspection and testing have been carried out in accordance with BS 7671, as amended. Cables concealed within trunking and conduits, or cables and conduits concealed under floors, in inaccessible roof spaces and generally within the fabric of the building or underground, have not been visually inspected.

* This Domestic Periodic Inspection Report should be reviewed and confirmed by the registered Qualified Supervisor of the Domestic Installer responsible for issuing it.

This form is based on the model Periodic Inspection Report shown in Appendix 6 of BS 7671: 2001. Published by NICEIC Group Limited © Copyright Electrical Safety Council (July 2006)

DPP1/1

NICEIC
DOMESTIC INSTALLER

G OBSERVATIONS AND RECOMMENDATIONS FOR ACTIONS TO BE TAKEN

Referring to the attached schedules of inspection and test results, and subject to the limitations at D:

There are no items adversely affecting electrical safety. □

or

The following observations and recommendations are made. ✓

Item No		Code †
1	EARTHING CONDUCTOR NOT CORRECT SIZE	2
2	SUPPLEMENTARY BONDING REQUIRED IN BATHROOM	2
3	RCD PROTECTION IS ADVISED FOR RING CIRCUITS	4

† Where observations are made, the inspector will have entered one of the following codes against each observation to indicate the action (if any) recommended:-

1. 'requires urgent attention' or
2. 'requires improvement' or
3. 'requires further investigation' or
4. 'does not comply with BS 7671:2001 (as amended)'

Note: If necessary, continue on additional page(s), which must be identified by the Domestic Periodic Inspection Report serial number and page number(s).

Please see the reverse of this page for guidance regarding the recommendations.

Urgent remedial work recommended for Items: □   Corrective action(s) recommended for Items: 1,2,3

H SUMMARY OF THE INSPECTION

General condition of the installation:

Note: If necessary, continue on additional page(s), which must be identified by the Domestic Periodic Inspection Report serial number and page number(s).

Date(s) of the inspection: 3-6-07

Overall assessment of the installation: UNSATISFACTORY

(Entry should read either 'Satisfactory' or 'Unsatisfactory')

I NEXT INSPECTION

I/We recommend that this installation is further inspected and tested after an interval of not more than: 10 YEARS

(Enter interval in terms of years or months, as appropriate)

provided that any items at G which have been attributed a Recommendation Code 1 (requires urgent attention) are remedied without delay. Items which have been attributed a Recommendation Code 2 or 3 should be actioned as soon as practicable (see G).

J SUPPLY CHARACTERISTICS, EARTHING AND BONDING ARRANGEMENTS   *Enter details, as appropriate*

Supply Characteristics

Nominal U/U_0 voltage:	230 V
Nominal frequency f [1]	50 Hz
Prospective fault current, I_{pf} [2]	0·9 kA
External earth fault loop impedance, Z_e [3]	0·58 Ω

Notes:
(1) by enquiry
(2) by enquiry or by measurement
(3) by enquiry or by measurement

System Type(s)

TN-S	✓
TN-C-S	
TT	

Characteristics of Primary Supply Overcurrent Protective Device(s)

BS(EN)	1361
Type	1
Nominal current rating	80 A
Short-circuit capacity	16·5 kA

Main Switch or Circuit-Breaker

Type: BS(EN)	5773
No of Poles	2
Supply conductors material	Cu
Supply conductors csa	16 mm²
Voltage rating	240 V
Current rating, I_n	100 A
RCD operating current, $I_{Δn}$	N/A mA
RCD operating time (at $I_{Δn}$)*	N/A ms

* (applicable only where an RCD is used as a main circuit-breaker)

Means of Earthing

Distributor's facility:	✓
Installation earth electrode:	N/A
Type: (eg rod(s), tape etc)	N/A
Electrode resistance, R_A:	N/A (Ω)
Location:	N/A
Method of measurement:	N/A

Main Protective Conductors

Earthing conductor		Main equipotential bonding conductors	
Conductor material	Cu	Conductor material	Cu
Conductor csa	6 mm²	Conductor csa	10 mm²
Continuity check	✓ (✓)	Continuity check	✓ (✓)

Bonding of extraneous-conductive-parts (✓)

Water service	✓	Gas service	N/A
Oil service	N/A	Structural steel	N/A
Location (where not obvious)		Lightning protection	N/A
		Other incoming service(s)	N/A

Please see the 'Notes for Recipients' on the reverse of this page.

Page 2 of 3

This form is based on the model Periodic Inspection Report shown in Appendix 6 of BS 7671: 2001. Published by NICEIC Group Limited © Copyright Electrical Safety Council (July 2006)

K SCHEDULE OF ITEMS INSPECTED
† See note below

Methods of protection against electric shock

Prevention of mutual detrimental influence

Item	Mark
Insulation of live parts, and barriers or enclosures	✓
Presence of RCD(s) for supplementary protection against direct contact and/or protection against indirect contact	N/A
Presence of earthing conductor and circuit protective conductors	✓
Presence of main equipotential bonding conductors	✗
Presence of supplementary equipotential bonding conductors	N/A
Class II fixed equipment	
SELV	N/A

Prevention of mutual detrimental influence

Item	Mark
Proximity of non-electrical services and other influences	✓
Segregation of Band I and Band II circuits or Band II insulation used	✓
Electrical separation	✓

Identification

Item	Mark
Presence of diagrams, instructions, circuit charts and similar information	✓
Presence of danger notices	✓
Presence of other warning notices	✓

Identification (cont)

Item	Mark
Labelling of protective devices, switches and terminals	✓
Identification of conductors	✗

Cables and conductors

Routing of cables in prescribed zones or within mechanical protection	✓
Connection of conductors	✓
Erection methods	✓
Selection of conductors for current carrying capacity and voltage drop	✓
Presence of fire barriers, suitable seals and protection against thermal effects	✓

General

Item	Mark
Presence and correct location of appropriate devices for isolation and switching	✓
Adequacy of access to switchgear and other equipment	✗
Particular protective measures for special installations and locations	✓
Connection of single-pole devices for protection or switching in phase conductors only	✓
Correct connection of accessories and equipment	✓
Choice and setting of protective and monitoring devices (for protection against indirect contact and/or overcurrent)	✓
Selection of equipment and protective measures appropriate to external influences	✓
Selection of appropriate functional switching devices	✓

L SCHEDULE OF ITEMS TESTED
† See note below

External earth fault loop impedance, Z_e	N/A
Installation earth electrode resistance, R_A	N/A
Continuity of protective conductors	✓
Continuity of ring final circuit conductors	✓
Insulation resistance between live conductors	✓
Insulation resistance between live conductors and earth	✓
Polarity	✓
Earth fault loop impedance, Z_s	✓
Operation of residual current device(s)	N/A
Functional testing of assemblies	✓

M CIRCUIT DETAILS

Circuit number	Circuit designation	Type of wiring (see code)	Reference method 4 of BS 7671 (see Appendix 4)	Number of points served	Live (mm²)	cpc (mm²)	Max. disconnection time permitted by BS 7671 (s)	BS (EN)	Type	Rating (A)	Short-Circuit capacity (kA)	RCD Operating current $I_{Δn}$ (mA)	Maximum Z_s permitted by BS 7671
1	COOKER	A	6	1	6	2.5		3036	5	30	1		N/A 1.14
2	RING 1	A	6	10	2.5	1.50		3036		30	1		N/A 1.14
3	RING 2	A	6	8	2.5	1.50		3036		30	1		N/A 1.14
4	IMMERSION HEATER	A	6	1	1.5	1.0		3036		15	1		N/A 2.67
5	LIGHTING	A	6	8	1.0	1.0		3036		5	1		N/A 10
6	LIGHTING	A	6	9	1.0	1.0		3036		5	1		N/A 10
7													
8													
9													
10													
11													
12													

N TEST RESULTS

	Ring final circuits only (measured end to end)			All circuits (At least one column to be completed)		Insulation resistance			Polarity	Maximum measured earth fault loop impedance, Z_s	RCD operating times		
	r_1 (Phase)	r_n (Neutral)	r_2 (cpc)	$R_1 + R_2$	R_2	Phase/Neutral (MΩ)	Phase/Earth (MΩ)	Neutral/Earth (MΩ)	(✓)	(Ω)	at $I_{Δn}$ (ms)	at 5 $I_{Δn}$ (if applicable) (ms)	
				0·24		>200	>200	>200	✓	0·82			
	0·35	0·35	0·57	0·23		>200	>200	>200	✓	0·81			
	0·39	0·40	0·64	0·24		>200	>200	>200	✓	0·82			
				0·23		>200	>200	>200	✓	0·81			
				0·91		>200	>200	>200	✓	1·49			
				1·31		>200	>200	>200	✓	1·95			

Test instruments (serial numbers) used:

Insulation resistance	CJK1041
Earth fault loop impedance	CJK1041
RCD	CJK1041
Earth electrode resistance	CJK1041

CODES FOR TYPE OF WIRING

A	PVC/PVC cables
B	PVC cables in metallic conduit
C	PVC cables in non-metallic conduit
D	PVC cables in metallic trunking
E	PVC cables in non-metallic trunking
F	PVC/SWA cables
G	XLPE/SWA cables
H	Mineral-insulated cables
O	Other - please state

† All boxes must be completed. '✓' indicates that an inspection or a test was carried out and that the result was satisfactory. '✗' indicates that an inspection or a test was unsatisfactory. 'N/A' indicates that an inspection or a test was not applicable to the particular installation. 'LIM' indicates that, exceptionally, a limitation agreed with the person ordering the work (as recorded in Section D) prevented the inspection or test being carried out.

This form is based on the model Periodic Inspection Report shown in Appendix 6 of BS 7671: 2001. Published by NICEIC Group Limited © Copyright Electrical Safety Council (July 2006)

NICEIC DOMESTIC INSTALLER

SCHEDULE OF CIRCUIT DETAILS
FOR THE INSTALLATION

TO BE COMPLETED IN EVERY CASE	ONLY TO BE COMPLETED IF THE DISTRIBUTION BOARD IS NOT CONNECTED DIRECTLY TO THE ORIGIN OF THE INSTALLATION*

Location of distribution board:

Distribution board designation:

Supply to distribution board is from:

Overcurrent protective device for the distribution circuit:
Type: BS(EN)
Rating: A

No of phases:
Nominal voltage: V

Associated RCD (if any): BS(EN)
RCD No of poles:
$I_{\Delta n}$ mA

CIRCUIT DETAILS

Circuit number and phase	Circuit designation	Type of wiring (see code below)	Reference method	Number of points served	Circuit conductors: csa Live (mm²)	cpc (mm²)	Max. disconnection time permitted by BS 7671 (s)	Overcurrent protective devices BS (EN)	Type No	Rating (A)	Short-circuit capacity (kA)	RCD Operating current $I_{\Delta n}$ (mA)	Maximum Z_s permitted by BS 7671 (Ω)
1	SHOWER	A	6	1	6	2·5	5	60898	B	32	3	N/A	1·5
2	RING	A	6		4	1·50	0·4	60898	B	32	3	N/A	1·5
3	RING	A	6		2·5	1·50	0·4	60898	B	32	3	N/A	1·5
4	IMMERSION	A	6	1	2·5	1·5	5	60898	B	16	3	N/A	2·4
5	LIGHTING	A	6		1·5	1·0	5	60898	B	6	3	N/A	8
6	LIGHTING	A	6		1·0	1·0	5	60898	B	6	3	N/A	8

† See Table 4A1 of Appendix 4 of BS 7671.

CODES FOR TYPE OF WIRING								
A	B	C	D	E	F	G	H	O (Other - please state)
PVC/PVC cables	PVC cables in metallic conduit	PVC cables in non-metallic conduit	PVC cables in metallic trunking	PVC cables in non-metallic trunking	PVC/SWA cables	XLPE/SWA cables	Mineral-insulated cables	

Page 4 of

* In such cases, details of the distribution (sub-main) circuit(s), together with the test results for the circuit(s), must also be provided on continuation schedules.

This form is based on the model shown in Appendix 6 of BS 7671: 2001.
© Copyright Electrical Safety Council (July 2006).

See next page for Schedule of Test Results

ICM3/7

SCHEDULE OF TEST RESULTS
FOR THE INSTALLATION

ONLY TO BE COMPLETED IF THE DISTRIBUTION BOARD IS NOT CONNECTED DIRECTLY TO THE ORIGIN OF THE INSTALLATION	Test instruments (serial numbers) used:	
★ See note below — Characteristics at this distribution board	Earth fault loop impedance	RCD
Z_s ★ — Ω — Operating times of associated RCD (if any) — At $I_{\Delta n}$ — ms	Insulation resistance	Other
I_{pf} ★ — kA — At $5 I_{\Delta n}$ (if applicable) — ms	Continuity	Other

TEST RESULTS

Circuit number and phase	Circuit impedances (Ω)					Insulation resistance † Record lower or lowest value				Polarity	Maximum measured earth fault loop impedance, Z_s ★ See note below	RCD operating times	
	Ring final circuits only (measured end to end)			All circuits (At least one column to be completed)		Phase/Phase †	Phases/Neutral †	Phases/Earth †	Neutral/Earth			at $I_{\Delta n}$	at $5 I_{\Delta n}$ (if applicable)
	r_1 (Phase)	r_n (Neutral)	r_2 (cpc)	$R_1 + R_2$	R_2	(MΩ)	(MΩ)	(MΩ)	(MΩ)	(✓)	(Ω)	(ms)	(ms)
1				0·14							1·6		
2	0·53	0·53	0·87	0·35							0·78		
3	0·5	0·5	0·82	0·33							0·76		
4				0·33							0·76		
5				0·96							1·39		
6				1·04							1·47		

★ Note: Where the installation can be supplied by more than one source, such as a primary source (eg public supply) and a secondary source (eg standby generator), the higher or highest values must be recorded.

TESTED BY

Signature:		Position:		Page 5 of [ ]
Name: (CAPITALS)		Date of testing:		

This form is based on the model shown in Appendix 6 of BS 7671: 2001.
© Copyright Electrical Safety Council (July 2006).

See previous page for Circuit Details

ICM3/9

Answers to Questions

1. Any values which are indicated as greater than can be disregarded as the true value is unknown.

$$\frac{1}{5.6} + \frac{1}{8.7} + \frac{1}{12} + \frac{1}{7} = 0.51 \qquad \frac{1}{0.51} = 1.92 \text{ M}\Omega$$

Enter into calculator as:

$$5.6 \, \text{x}^{-1} + 8.7 \, \text{x}^{-1} + 12 \, \text{x}^{-1} + 7 \, \text{x}^{-1} = \text{x}^{-1} = 1.92$$

This is acceptable as the total insulation resistance is greater than 0.5 MΩ and each circuit is greater than 2 MΩ.

2. A 2.5 mm^2/1.5 mm^2 cable has a resistance of 19.51 mΩ per metre. The resistance of 54 metres is:

$$\frac{54 \times 19.51}{1000} = 1.05 \ \Omega \qquad \frac{1.05}{4} \ 0.26 \ \Omega$$

$R_1 + R_2$ for the circuit is 0.26 Ω

$$Z_S = Z_e + R_1 + R_2 \qquad 0.24 + 0.26 = 0.5 \ \Omega$$

From Table 41B2 in BS 7671, the Maximum value for a type C32 amp device is 0.75. Use the rule-of-thumb to compensate for conductor operating temperature and ambient temperature.

$$0.75 \times 0.75 = 0.56 \, \Omega$$

Therefore this circuit will comply.

3. From Table 9A in the *On-Site Guide*, the $R_1 + R_2$ value for the $4 \, \text{mm}^2/1.5 \, \text{mm}^2$ conductors is $16.71 \, \text{m}\Omega$ /M. The $R_1 + R_2$ value at the sockets is:

$$\text{Socket 1:} \quad \frac{12 \times 16.71}{1000} = 0.2 \, \Omega$$

$$\text{Socket 2:} \quad \frac{18 \times 16.71}{1000} = 0.3 \, \Omega$$

$$\text{Socket 3:} \quad \frac{20.5 \times 16.71}{1000} = 0.34 \, \Omega$$

Z_S will be $0.34 + 0.6 = 0.94 \, \Omega$. The maximum Z_S for a 30A BS 3036 fuse from Table 41B1 in BS 7671 (0.4s) is given as $1.14 \, \Omega$.

Using the rule-of-thumb for temperature correction the maximum permissible value is: $1.14 \times 0.75 = 0.86 \, \Omega$. This value is acceptable.

4. From Table 9A in the *On-Site Guide*, the $R_1 + R_2$ value for $10 \, \text{mm}^2/4 \, \text{mm}^2$ copper is $6.44 \, \text{m} \, \Omega/\text{M}$.

$$R_1 + R_2: \quad \frac{6.44 \times 14.75}{1000} = 0.095 \, \Omega$$

Z_S for the circuit is $0.095 + 0.7 = 0.795 \, (0.8) \, \Omega$.

The maximum Z_S for a 45A BS 3036 fuse from Table 41D in BS 7671 (5s) is $1.66 \, \Omega$.

Corrected for temperature: $1.66 \times 0.75 = 1.24 \, \Omega$. As the actual Z_S is lower the circuit will comply.

5. (i) From Table 9A in the *On-Site Guide* $4 \, \text{mm}^2$ copper conductors:

have a resistance of $4.61 \, \text{m}\Omega/\text{M}$.

$R_1 + R_2$ for the phase and CPC both in $4 \, \text{mm}^2$ is $9.22 \, \text{m}\Omega/\text{M}$

$$R_1 + R_2: \quad \frac{9.22 \times 67}{1000} = 0.061 \, \Omega$$

As it is a ring:

$$\frac{0.61}{4} = 0.15 \, \Omega$$

The $R_1 + R_2$ value is $0.15 \, \Omega$.

$$Z_S = Z_e + R_1 + R_2$$

$0.63 + 0.15 = 0.78 \, \Omega$ as the maximum permissible is given as $0.86 \, \Omega$ this value is acceptable.

(ii) As the maximum permissible Z_S is given as 0.86 and is taken from the *On-Site Guide*. No correction for temperature is required. We must subtract the Z_e from the actual Z_S to find the maximum permissible $R_1 + R_2$ value.

$$R_1 + R_2 = 0.86 - 0.63 = 0.23 \, \Omega$$

We must now subtract the actual $R_1 + R_2$ value from the maximum permissible value.

$$0.23 - 0.15 = 0.08 \, \Omega$$

The maximum resistance that our spur could have is 0.08Ω. To calculate the length we must transpose the calculation:

$$\frac{\text{mV} \times \text{length}}{1000} = R$$

to find the total length transpose to

$$\frac{R \times 1000}{mV} = \text{length}$$

Therefore:

$$\text{length} = \frac{0.08 \times 1000}{9.22} = 8.67 \text{ metres}$$

Total length of cable for the spur will be 8.67 metres.

6. Electrical installation certificate
   Schedule of test results
   Schedule of inspection.

7. BS 7671 Wiring regulations
   On-Site Guide
   Guidance note 3
   GS 38.

8. Due date
   Clients request
   Change of use
   Change of ownership
   Before alterations are carried out
   After damage such as fire or overloading
   Insurance purposes.

9. Where there is recorded regular maintenance.

10. (i) P-N are $0.3\,\Omega$ each. The total loop will be 0.6 and the P-N at each socket after interconnection will be:

$$\frac{0.6}{4} = 0.15 \text{ ohms}$$

(ii) CPC must be $0.3 \times 1.67 = 0.5\,\Omega$

(iii) $R_1 + R_2$ loop will be $0.5 + 0.3 = 0.8\,\Omega$. after interconnection the $R_1 + R_2$ value at each socket on the ring will be:

$$\frac{0.6}{4} = 0.2$$

11. $2.5\,\text{mm}^2/1.5\,\text{mm}^2$ has a resistance of $19.51\,\text{m}\Omega$ per metre. 5.8 metres of the cable will have a resistance of:

$$\frac{5.8 \times 19.51}{1000} = 0.113$$

0.113 is the resistance of the additional cable. $R_1 + R_2$ for this circuit will now be $0.2 + 0.113 = 0.313\,\Omega$.

12. A statutory document is a legal requirement.

13. A non-statutory document is a recommendation.

14. Safety as the satisfactory dead tests ensure that the installation is safe to energize.

15. There are nine special locations.

16. The insulation resistance would decrease.

17.

$$\frac{1}{50} + \frac{1}{80} + \frac{1}{60} + \frac{1}{50} = 0.069 \qquad R = \frac{1}{0.069} = 14.45\,\Omega$$

18. Continuity of bonding and CPCs
    Ring final circuit test
    Insulation resistance test
    Polarity
    Live polarity at supply
    Earth electrode (Z_e)
    Prospective short circuit current
    Residual current device
    Functional tests.

19. Low resistance ohm meter
    Low resistance ohm meter
    Insulation resistance tester
    Low resistance ohm meter
    Approved voltage indicator
    Earth loop impedance meter
    Prospective short circuit current tester
    RCD tester.

20. 15 mA, 30 mA, 150 mA (*only 150 mA if used for supplementary protection*).

21. Five times.

22. 0.05 Ω.

23. From Table 9A in the *On-Site Guide*, 10 mm^2 copper has a resistance of 1.83 mΩ per metre:

$$\frac{1.83 \times 22}{1000} = 0.4 \ \Omega$$

24. TT system.

25. Socket 1: Good circuit
    Socket 2: CPC and N reversed polarity
    Socket 3: P and N reversed polarity
    Socket 4: P and CPC reversed polarity
    Socket 5: Spur
    Socket 6: Loose connection of N
    Socket 7: Good circuit.

26. From Table 41D Z_S for a 5A BS 3036 fuse is 18.5 Ω.
    The $R_1 + R_2$ value for 1 mm^2 copper from Table 9A is 36.2 mΩ.
    Maximum resistance permissible for the cable

$$18.5 - 0.45 = 18.05 \ \Omega$$

Maximum length of circuit is:

$$\frac{18.05 \times 1000}{36.2} = 498 \ \text{M}$$

(*Problem with volt drop if the circuit was this long*).

27. TT: 21 Ω
    TNS: 0.8 Ω
    TNCS: 0.35 Ω

28. Twelve socket outlets, one for each socket on the ring.

29. Unlimited number.

30. GS 38.

31. Taps, Radiators, Steel bath. Water and gas pipes, etc.

32. Steel conduit and trunking. Metal switch plates and sockets. Motor case, etc.

33. $4\,\text{mm}^2$.

34. $0.5\,\text{M}\Omega$.

35. 500 volts, 1 mA.

36. In possession of technical knowledge or experience or suitably supervised.

37. 500 Volts DC.

38. Flexible. Long enough but not so long that they would be clumsy. Insulated. Identified. Suitable for the current.

39. Finger guards. Fused. Maximum 4 mm exposed tips. Identified.

40. Approved voltage indicator or test lamp.

41. Fixed equipment complies with British standards, all parts correctly selected and erected, not damaged.

42. To ensure no danger to persons and livestock and that no damage occurs to property. To compare the results with the design criteria. Take a view on the condition of the installation and advise on any remedial works required. In the event of a dangerous situation, to make an immediate recommendation to the client to isolate the defective part.

43. Ensure the fault is repaired and retest any parts of the installation which test results may have been affected by the fault.

44. To ensure that all single pole switches are in the phase conductor. Protective devices are in the phase conductor. ES lampholders are correctly connected. The correct connection of equipment.

45. E14 and E27 as they are all insulated.

46. The top surface must comply with IP4X. The sides and Front IP2X or IPXXB.

47. By calculation $Z_S = Z_e + R_1 + R_2$. Or use a low current test instrument.

48. 0.37 kW (Regulation 552-01-01).

49. (a) Type C
(b) Type D
(c) Type C

50. Safety electrical connection do not remove. Voltage in excess of 230 volts where not expected. Notice for RCD testing. Where isolation is not possible by the use of a single device. Where different nominal voltages exist. Periodic test date. Warning non-standard colours.

51. $1.2 \times 0.75 = 0.9\,\Omega$
$0.96 \times 0.75 = 0.72\,\Omega$
$1.5 \times 0.75 = 1.125\,\Omega$
$4 \times 0.75 = 3\,\Omega$
$2.4 \times 0.75 = 1.8\,\Omega$

52. IP 4X

53. $10\,\text{mm}^2$.

54. Electrical installation certificate
Schedule of test results
Schedule of inspection

55. BS 7671 (Electrical Wiring Regulations)
On-Site Guide
GS 38
Guidance note 3

56. Health and Safety at Work Act 1974
Electricity Supply Regulations
Electricity at Work Regulations 1989
Construction Design and Management Regulations
Building Regulation Part P
(Appendix 2 of BS 7671 covers statutory regulations).

57. Brown (L1), Black (L2), Grey (L3) and Blue (N).

Appendix

F

Useful information

The City and Guilds 2391 inspection and test exam is based on information given in Guidance note 3, BS 7671 Wiring Regulations, the *On-Site Guide* and GS 38. It is important that any person intending to take this exam has a sound knowledge of these documents.

The 2391 101 written examination consists of two sections. The first section (A) contains 20 short answer questions. Of these 20 questions 15 will be on testing, three will be on preparation of testing and two will be on the process of testing. Each question will be worth 3 marks. The question may be split into parts each worth 1 mark or it may require a single answer worth 3 marks.

The second section (B) will consist of six questions, each worth 15 marks, these are long answer questions and are based on a scenario which is given at the beginning of the section. Within each of the questions marks are given for each correct part of the answer.

Section B of the paper will require in-depth answers; for example, the question could ask how you would carry out an insulation test on a domestic lighting circuit. It is easier to answer the question using bullet points as shown than it is to write an essay.

- Ensure safety whilst test is being carried out.
- Isolate circuit to be tested using the safe isolation procedure.
- Remove all lamps.

- If the removal of lamps is difficult or fluorescent/transformers are on the circuit open the switches controlling them.
- Isolate or bypass all electronic equipment and equipment that may be damaged or give false readings.
- Use an insulation resistance tester set on 500 V d.c.
- Ensure instrument is accurate and operate the instrument with the leads together and then apart to ensure correct operation.
- Test circuit between live conductors.
- Operate any two-way switches controlling points where lamps have been removed.
- Join live conductors together and test between them and earth.
- Operate all two way switches.
- The insulation resistance should be above $2\,M\Omega$. If it is less, then further investigation must be carried out as a latent defect may exist.
- An insulation resistance value of $0.5\,M\Omega$ is acceptable in some cases.
- Replace all lamps and remove any shorting links used.
- Leave circuit safe and operational.

In any exam it is vital that the question is read carefully. Often it is better to read a question several times to try and understand what is being asked.

It is always better to show all calculations even if the question does not require you to. It is a good habit to get into and often you will get marks for showing the correct calculation even if the answer is wrong.

If the question asks for a fully labelled diagram, then marks are awarded for the diagram and the labelling. Where the question asks 'explain with the aid of a diagram' then a diagram and a written explanation is required.

When the question asks for a list then you will be expected to list a sequence of events in the correct order. *Just a list with no explanation.* If you are required to state something then a statement is required in no particular order.

For example:

List the sequence of dead tests.

a. Continuity of bonding conductors and circuit protective conductors.
b. Ring final circuit test.
c. Insulation resistance test.
d. Polarity.

State three statutory documents relating the inspecting and testing of electrical installations.

- The Electricity at Work Regulations 1989
- The Health and Safety at Work Act 1974
- The Electricity Supply Regulations.

Always try to answer the questions in full using the correct terminology; for example: if asked "which is the type of inspection to be carried out on a new installation", the answer must be: An initial verification. For the document required for moving a switch or adding a socket the answer must be: An electrical installation minor works certificate, not just minor works.

Do not waste time copying out the question and write as clearly as you can.

Index